BEER O'CLOCK

How to create beer, make friends and lose
inhibitions while sitting around the house

Written by Master Brewer, Geoff Wallace

ISBN 978-0-6484277-0-4

BEER O'CLOCK
COPYRIGHT 2018 Michael Wallace

"Instead of water we got here a draught of beer…a lumberer's drink, which would acclimate and naturalize a man at once—which would make him see green, and, if he slept, dream that he heard the wind sough among the pines."
Henry Thoreau

In Memoriam: Geoffrey John Wallace
Master Brewer and Author

Passed On August 13th 2016

"I would doubt there is a more concise, detailed and knowledgeable book on the subject of beer making than this book. It is packed with the essential information and backed with experience in the practical day-to-day matters of brewing"

"Fill with mingled cream and amber,
I will drain that glass again.
Such hilarious visions clamber
Through the chambers of my brain.
Quaintest thoughts — queerest fancies,
Come to life and fade away:
What care I how time advances?
I am drinking ale today."

Edgar Alan Poe

BEER O'CLOCK!

How to create beer, make friends and lose
inhibitions while sitting around the house

Geoff Aged Fifty Five, at the
height of his brewing prowess

Forty Years Experience in Brewing
By: G J Wallace

BEER Wonderful BEER

Have you ever wondered about BEER? It is the ubiquitous beverage consumed in various forms all over the world, with a heritage dating back Thirteen thousand years or more. In every continent, in every culture, there is a form of BEER.

Evidence suggests that stone mortars from Raqefet Cave, Israel, were used in brewing cereal-based beer millennia before the establishment of sedentary villages and cereal agriculture.

In Mesopotamia, there is an almost four thousand year old poem dedicated to the Goddess of BEER, one that just happens to contain the recipe for making it. The FIRST book on how to brew BEER!

We believe the Bavarians invented Lager in the 15th Century, but there is evidence on pots from one thousand years ago that LAGER YEAST (Saccharomyces eubayanus) was being used in Patagonia, in South America.

It is a wondrous product, and some say BEER is the precursor to civilization as we know it, as it formed a REASON to cultivate grains beyond mere bread making.

This book has been compiled from work done by my father some thirty years ago. You could say he was a devout believer in beer and the good it can do you. I know it made his very difficult life bearable. And so, to celebrate his life I give you the book he called "that rubbish" but which just happens to be one of the best books on how to brew beer that you will ever find.

Dedicated to Geoffrey John Wallace, a loving father, a good friend, and an excellent maker of booze.

"I have a deep gratitude for YEAST, without which we would have no BEER."

Introduction

Geoffrey John Wallace was an extraordinary brewer. He could taste a beer, any beer, and by magically mixing a few ingredients, he could copy it so well that few could tell the difference between the 'professional' product and his home brew. He did this for close to forty years, but even before beer, he made a small fortune making and selling 'pop' fizzy drink in New Guinea.

He loved brewing. It meant he was CREATING something of value. I still smile when recalling how he would wait in anticipation for the 'new brew' he had made would be ready. He would crack it open (never before Five PM - BEER O'CLOCK - unless it was the weekend) pour some into a glass, and hold it up to the light like a wine connoisseur. Geoff at seventy years of age still had the eager look of a child about to open a wonderful present. Then he would taste it and thoughtfully swish it about before swallowing. You could see the gears running. Was it too tart, or too sweet? Was there enough Malt? Was it crisp and clean?

He would then go to his BREWERS NOTEBOOK and write down his impression of how a particular batch has turned out. Only then would he appear satisfied and relax to drink the beer.

Geoff Wallace was a meticulous note taker. Every tiny detail of every brew was recorded. Let me not mention the 'other' notebook with everything to do with the car or the notebook that had everything to do with the bills, etc. The point is, he kept accurate and detailed records, and because of this, he can speak with AUTHORITY, not just from personal opinion. Everything recorded in this book derives from experience and cold hard facts.

He was an enthusiastic tester of his own product, of which there was a significant volume. Under the block of flats he had at 502 Milton Road, Brisbane, there would be up to and over two thousand two hundred bottles, all full of his sparkling ales and stouts, at any given point in time. On occasions, late at night when the party where I was ran dry, a dozen or so would be appropriated, with a $20 bill left behind as a thank you. I mentioned this late in his life, but it appeared he never really noticed.

Beer as Proof Of God

I would often watch as he would prepare all the ingredients from scratch. He malted his own barley, activated his own hops, and while I watched he would describe at every step of the way precisely what had to be done to get the perfect result. By way of his humor, after he had detailed every tiny step, Geoff would say "And this proves the existence of God!"

Obviously, the first time I asked why, and he responded, "Because the malt must be held at this specific temperature for that length of time, and the hops must be dry roasted at THIS particular temperature for THAT specific length of time. Unlike wine, you could not 'stumble' onto how to make beer. Everything about it is designed and scientific. And as primitive cultures didn't have thermostats or clocks or hydrometers, clearly someone GAVE them the secret. It was either aliens, or God, and I don't believe in aliens!"

This sort of tongue-in-cheek humor was legendary with my father, yet in all his quips he would hide a kernel of truth that he wanted you to catch. He had a deep wisdom in regards people and he often said: "Humour is the best lubricant between the gears of human society."

I had thought this book long lost, but two years after his passing (Aged 91 with all his faculties still intact. A good advertisement for home brew!) I found this manuscript in a box, in a shed, where I had stored some of his stuff.

I had asked about it many times when he was alive, but he always waved me aside, saying, "No one wants to read that rubbish." It was typical of his humility that what I considered to be the best and most articulate book ever written on how to create home brew beer he referred to as 'that rubbish'.

I found it both sad and yet enjoyable to go over his hand-typed notes, thumped out the translucent foolscap that he always used when typing. On every sheet, I could see the physical corrections, the whiteout, and the areas where he has changed his mind about this or that. But overall, there had been precious few alterations to the original text.

Accordingly, this book is largely as he wrote it, with minor redactions, such as the pricing of items he gave in the original, now long out-of-date, manuscript. Another issue was that he converted Metric to Imperial Gallons, etc. rather than US Gallons. At the time of writing, in the 1980's, there were a lot of people in Australia who still thought in terms of British Imperial measurements. I don't imagine he ever even dreamed his book might reach the Americas, but of course, Amazon has changed all this. If I have made mistakes in conversions or missed some, forgive me. For the record, 5 IMPERIAL Gallons = 5.93 US Gallons. Almost all Fermenters in his day were 5 Imperial Gallons.

Geoff's Violent Upbringing

Geoff Wallace brewed his own beers not just because he loved to drink (he was a very happy drunk), he brewed his own beer because it was fun and because he made a better product than you could buy, not just because it was cheaper.

He came from an incredibly difficult and stressful upbringing. For Example: His father would point a loaded shotgun at his head, showing him it was loaded and with the safety off, then give him spelling tests. The father had returned from the trenches of WW1 a violent and dangerous man. Geoff generally referred to him (Duncan Harry Wallace) as "that bastard". But he did not pass this anger on to his family, rather he contained it and drank excessively to control the stress.

Deep down, underneath everything, I suspect the reason he (and his brothers) drank was to deal with the hurt from their childhood. But for Geoff it went further: His father was not just brutal, his mother effectively traded him (her youngest born) in order to get the insane husband out of her life. This hurt stayed with him his whole life, but he said nothing to anyone but his brother. I only discovered the secret regarding his mother after he passed on, in a letter to his last remaining sibling.

Geoff did not indulge in the mood swings or the verbal and physical assaults he experienced from his father. I only saw kindness and generosity of spirit from him, something completely lacking in his own childhood. Largely because of this I had long wanted to publish this book to honor his life, as a small way of saying 'thank you'. So it is was with a great deal of joy that I discovered it in storage. At last I can give you his simple masterpiece: Beer O'Clock.

If, at the end of this book, you wish to know more of Geoffrey John Wallace, I heartily recommend to you the biography I wrote of him for the family, "The Parables of Geoff". (Available on Amazon - goo.gl/mGSHwn)

Please note: I have optimized the book for American Spelling, color instead of colour, liter instead of a litre, etc. If you are from a non-US spelling country, please overlook this. It would be appreciated. I simply had to aim for the US market as the most likely and largest area of interest for his book.

One of Geoff's favorite photos. As he would say, "You would go out of your way to drink, just so they wouldn't want to kiss you!"

INDEX

7 INTRODUCTION
11 What is Beer?
12 The "Easy Way" method
13 Brewers Notebook
14 The "Easy Way" Economical Recipe
16 Do I use the "Easy Way" or a Beer Pack?
18 Equipment you will need
19 How the Beer Making equipment is used.
20 Deciding What Beer to Brew
22 (Testing Your Fermenter)
23 Making the Sterilising Solution
34 BREWING: Easy Step by Step Instructions
- Step One: Bring the water to boil
- Step Two: Add Ingredients
- Step Three: Putting your mix into the Fermenter
- Step Four: Pitching the Yeast
- Step Five: Fermentation
- Step Six: Topping up the Fermenter
- Step Seven: Allowing the Wort to settle
- Step Eight: Bottling
- Step Nine: Storing the Bottled Beer
- Step Ten: Recovering the Yeast from the Fermenter
- Step Eleven: Cleaning the Fermenter
36 **Trouble Shooting**
43 Bottles
45 Household Bleach - The Brewers Friend
47 Bottle Washing
48 Yeast
50 Adding Finings
51 The Hydrometer
54 How to Calculate the Potential Alcohol Strength
55 How to Calculate Alcohol Strength when Water is to be Added
56 Note on World Proof Systems
57 Calculating the Amount of Water to Add to Bottles
58 Ready Reckoner Table
59 Are Glucose and Dextrose the same thing?
60 Pouring Home Brew
61 Filtration
62 Record Keeping
64 Experimenting
65 Cockroaches in the Wort
66 Beer as a Fire Extinguisher
67 How to Remove Bottle Sediment when Traveling
68 Types of Alcohol
69 Blood Alcohol Level in Drivers
70 Effects of Alcohol in the body
72 Acidity, Alkalinity and the PH Scale
73 Starter Bottles
74 Calculate Alcohol Based on the Sugars Used
76 Invert Sugar
77 The Thermometer
78 Open Fermenters
81 The Absolute importance of sterilization
82 Commercial Beer Making
83 Brewing Terms

Ancient Egyptian brewing Beer

What is BEER?

WELCOME! I trust you will not only enjoy this book but, as a result of reading it, you will get out there and practice the fine art of brewing. In here is everything you will need to understand the basics of making BEER. In this book you will find clear, foolproof instructions on how to make an excellent home brew beverage.

I suspect anyone reading this book has a good idea what Beer is, at least they believe they do. But I promise you, it is much more than you ever imagined and holds a far deeper science then you might have believed. Technically speaking, however, Beer consists of MALT, HOPS, WATER, and YEAST. That is IT. The varying amounts of these ingredients and the way these are treated determines the flavor and taste of your beer.

In some countries, the above are the *only permissible ingredients* allowed by law in commercial beers. But, of course, the home brewer is free to add whatever he chooses. For example, sugar and cereal grains may be added to increase alcohol content or to impart additional flavor and color. I will describe what each characteristic ingredient offers:

Malt gives flavor, color, and aroma. Being a fermentable sugar, Malt contributes to the alcohol content of the beer. The HOPS give the bitter taste and some aroma, and it is also a preservative. The WATER, obviously, is the quantity. The YEAST, by fermenting the mixture above, converts the SUGARS into ALCOHOL and CARBON DIOXIDE GAS.

With home-made beer, upon bottling, a teaspoon of sugar is added to each large sized bottle (750 ml or 26 fl oz) and this starts a secondary fermentation process, working with the carry-over yeast still in suspension in the beer. This stage produces a little more alcohol plus Carbon Dioxide gas, the same gas used to put the 'fizz' into soft drinks, or pop. sparkling wines and commercial beers. As the bottle is capped (sealed) at filling time, this gas cannot escape and so it dissolves into the beer, producing the many fine bubbles, life, and attractive foaming head when the beer is poured.

Sugar of one or more kinds is converted by the YEAST into ALCOHOL and loses all of its sweetness. Around four weeks, or less, after bottling (the longer the better, up to a point) the beer is ready to drink and enjoy. Two weeks is usually, but not always, the minimum time.

The basic process of making beer is quite simple, but it is meticulous. Care must be taken every step of the way to ensure a safe product, and one that passes the 'taste test'. Once you grasp the detailed instructions as outlined in this book, you will merely need a little experience to be making first-class beer that you and your friends can enjoy.

All you need remember is CLEANLINESS, STERILIZATION and CAREFUL NOTATION of everything you do and there are very few things that will go wrong.

The "Easy Way"

In this opening section, I outline a simple and effective technique that will get you up and brewing like a master in no time. It is called the "Easy Way" because this is exactly what it is. This chapter shows you just how simple it is, and later we will go into refinements.

You add MALT EXTRACT and SUGAR (or Dextrose) with either HOP PELLETS of HOP FLOWERS to water which is boiling in a large saucepan. *Simmer for 30 Minutes.* Then strain into a large 22.5 liter (5.9 US Gallons) container, called the FERMENTER. Add tap water to almost fill the Fermenter,. Add the Yeast from a packet and leave it for nine to ten days, by which time it is ready to bottle.

This produces thirty large bottles of beer. HOP EXTRACT may be substituted with HOP FLOWERS or HOP PELLETS. If so, it is sufficient to just bring your starter mix to the boil using only the MALT EXTRACT and the SUGAR. Remove this from the stove, then add the HOP EXTRACT to the Fermenter, (Not the hot fluid!) along with the water and yeast.

If you are using a BEER PACK, simply dissolve the contents into boiling water and pour into the Fermenter,. (All the other detailed instructions which follow still apply in full) If you have tried commercial beer packs, and the results have been poor, you will no doubt find the reason for this in this book, as most of the commercial offerings are very good.

Now, there is a point of dispute regarding the type of sugar to use. Some brewers prefer to use dextrose powder instead of sugar, as they claim the quality of the beer is improved. I am inclined to agree with them if taste is the only consideration. However, sugars developed from starch, such as dextrose, has some startling implications for health, and the development of fat in the liver. It is a matter of personal choice, however, white sugar is extremely inexpensive and is available everywhere. It does a perfectly good job. If you prefer dextrose and are happy to pay the price, then, by all means, use it.

To produce 30 bottles of beer takes very little time. If you are using the following recipe and choose HOP FLOWERS or HOP PELLETS to boil with the malt extract, it takes about forty minutes to have it all in the Fermenter. If using the recipe and HOP EXTRACT, it only takes five minutes to prepare the combination of elements for the fermentation vat. The EXTRACT is more expensive but if you don't mind the cost it is just as good as the others. Its advantage is that it doesn't have to be boiled or strained.

If you are using a COMMERCIAL BEER PACK (concentrated beer wort) it would take even less time, as this does not have to be brought to the boil. You can do without the saucepan and dissolve it in the Fermenter, but the saucepan is better.

Once the brew is in the Fermenter, there is not much to do until bottling time. The bottling process takes about thirty minutes, roughly one minute per bottle. So given all the above, and how easy the "Easy Way" is, I want to encourage you to give it a go as soon as possible. It is truly not much more difficult than making a cup of tea or coffee, as long as you have everything organized.

BREWER'S NOTEBOOK

I cannot stress enough the importance of detailing every little thing that you do. There is a good deal of time between the start of the process and the drinking of the beer at the end, and you WILL forget the small details of what you have done. Writing down the details means you have a 'map' of where you have been, so if anything goes wrong you can backtrack and discover where the trouble came from.

When you choose one of the following recipe's to start your first brew, write down exactly what you are adding to the WORT. Do NOT throw in, as an example, a cup of SUGAR and just say "Cup of Sugar". WEIGH IT FIRST, and write in 400 grams (14oz) Sugar, or whatever it might be. Not to be pedantic, but when using the scale you weight the cup first, then the cup with the sugar, and subtract the weight of the cup. Small details like this make an ENORMOUS difference regarding the quality of your end result.

When you take the Original Gravity (OG) of the WORT with your Hydrometer, WRITE IT DOWN. You WILL forget, or worse, after a few brews are under your belt you will remember a WRONG number. This WILL cause problems in the final outcome.

I don't want to appear like an old-fashioned school teacher, whacking the desk with a ruler, telling you what you must do, or else. But do it, or else.

May I suggest to you a simple step-by-step method? Start with the DATE, everything starts from that point. Write in the details of what mix you choose, with all ingredients weighed. Write down what HOPS you have chosen, the amount of MALT you use, and how much SUGAR you have added. Note how long to you took to boil it up. After you add it to the FERMENTER, note the YEAST you added, and if it is a batch from a former brew note what it was.

And then, every step of the way, on a single page (which is all you need) in your WORKBOOK, write down every detail. If you TOPPED UP on a specific day, note that and how long the lid was off the Fermenter. Every time you take a Hydrometer reading, note it down and the DATE you took it.

Even record when you STERILIZED your equipment because the vast majority of problems in home brewing come from a mistake in this area. You will only need a page in an A5 sized notebook. I had mine tied to the Brewing Table and wrapped in plastic, to make sure I never forgot it.

It will look a little like this:

10 June 1989. Prepared 1K MALT, 400g SUGAR, 40g HOP FOWERS and boiled for 40 minutes

Added to WORT with one packet of "x" brand Brewers Yeast

OG (Original Gravity) 1.039 WORT TEMP of 15.5C (60F)

15 June 1989: TOPPED UP. SG (Specific Gravity) 1.011

22 June 1989: FG (Final Gravity) 1.003, WORT TEMP 22C - ready to bottle

22 June 1989: BOTTLING: Added rounded teaspoon of WHITE SUGAR

31 Aug 1989, opened and tasted. Excellent brew. Could use a tad more malt.

"Easy Way" Recipes

The following comes from many years of experience and experimentation. These measurements are extremely accurate, and the alcohol content if you using exactly the ingredients listed per brew is quite accurate.

(Editor Note: When this book was compiled there was no such thing as the internet and people still used typewriters to record information. The entire book was written on a Remington Portable that Geoff had used for correspondence since post-war and his time in New Guinea. Currently, all the listed ingredients are easily available on EBay)

MEDIUM Strength Beer (around 3% alcohol)
1 Kg (2 lb 3 oz) DARK liquid malt extract
400g (14 oz) Sugar
40 g (1 1/2 oz) HOP FLOWERS
One sachet (packet) BREWERS YEAST

PRICE: Dark Malt Extract - 1 KG - around $13, Sugar 70 cents, HOP FLOWERS (40g) $6, Brewers Yeast - $1 = less than $21 for thirty bottles = 0.70 cents a bottle / 45 cents a stubby. You CAN save considerably on the above costs by buying in bulk.

LIGHT Beer - around 2.5% alcohol
MALT EXTRACT (as above)
200g (7oz) sugar
HOPS and YEAST (as above)

In other words - Simply use less sugar. Cost is much the same.

EXTRA LIGHT Beer - around 2% alcohol
MALT EXTRACT (as above)
NO SUGAR
HOPS and YEAST (as above)

Uses NO sugar thus alcohol converted from the malt alone. Cost is much the same.

Commercial Strength Beer (Around 4.25% alcohol)
1 Kg (2 lb 3 oz) DARK liquid malt extract
One Kilo (2lb 3oz) Sugar
40 g (1 1/2 oz) HOP FLOWERS
One sachet (packet) BREWERS YEAST

Overall cost around 72 cents a bottle

As you can easily see, the alcohol content is regulated by the amount of sugar used. You are not saving money brewing lower alcohol beer, but you can drink more of it before falling over, so it has advantages in a different direction than just cost.

All these recipes are for a 22.5 liter (5.9 US gallon) brew, which gives thirty large bottles. Prices are extremely variable, and buying larger quantities through wholesalers can reduce costs by half. Given that thirty large bottles represent 2.5 cartons of beer, or well over $100 in retail cost, you are making a home brew for approximately one fifth the cost.

Another way of seeing this is that for the same money you get FIVE TIMES more beer by brewing at home as opposed to going down to the bottle shop. And compared to bar prices at $5 plus for 375 mm, well, there's no real comparison.

A brief summary of measurements is as follows:
g = Gram
oz = Ounce (28 grams)
1lb = Pound (16 Oz or 453 grams)
kg - Kilogram or Kilo (1,000 grams or 2lb 3oz)
fl oz = Fluid Ounce (28 ml)

Standard Coopers Brew Kit. Contains everything you need.

"Stay with the beer. Beer is continuous blood. A continuous lover."
Bukowski

VARIATIONS: "Easy Way" Recipes

I f cost is not important and if it suits your taste, an extra 1/2 kilo of MALT EXTRACT can be added to each recipe. This gives a maltier flavor and will increase the alcohol content by around 3/4%. A lot of attention will be given in this book to alcohol content as drinkers are becoming more aware of the dangers of excess alcohol, and the need to regulate consumption if driving. This is evidenced by the increasing popularity of low-alcohol beers.

HOP PELLETS (mechanically concentrated hop flowers), when you can get them, are usually cheaper than HOP FLOWERS and give excellent results. Two heaped dessert spoons (50 gm, or a little under 2 oz) are used per brew, so they are also easier to handle. HOP EXTRACT is very good but more expensive. It is very concentrated and 5 ml per brew is sufficient.

HOP EXTRACT has the advantage of not needing to be boiled and can be added directly to the Fermenter. When using HOP EXTRACT is it sufficient to just bring the malt extract to the boil and then put it into the Fermenter with the HOP EXTRACT and the water. If you don't mind the added expense, HOP EXTRACT makes the entire process ridiculously easy, as it only takes minutes to prepare a brew.

DEXTROSE, a sweet white powder made from starch, is an often used substitute for sugar. It is around twice the price of sugar, and available on Ebay. Many brewers maintain that DEXTROSE produces a better beer. However, while DEXTROSE appears to be chemically identical to GLUCOSE, there is evidence that it produces significantly higher levels of fat in the liver than ordinary sugar. This book is not designed to educate you in biochemistry, nor am I qualified to do so, but it may be something you would like to research.

Informed choices make wiser decisions.

The aforementioned recipes can be extended to make 36 large bottles of beer (Three cartons) if desired by simply adding 128 ml (4 1/2 fl oz) of water to each bottle at bottling time. Some say it improves the flavor, I can't say I noticed a tremendous difference.

MALT EXTRACT is made in DARK and LIGHT grades. The light colored one is quite satisfactory if the dark in unavailable, but it doesn't impart enough color to the finished product. If you use it, add a dessert spoon or so of caramel color (Parisian Essence) but dissolve it in water first.

HOP FLOWERS should be boiled with the MALT EXTRACT and SUGAR for thirty minutes. HOP PELLETS can be boiled for a minimum of fifteen minutes with the MALT EXTRACT, or boiled separately and strained, then added to the Fermenter. HOP EXTRACT does not need boiling at all.

NOTE: Very detailed directions follow on later in this book.

If you are a beginner, the first "Easy Way" recipe is recommended. It has a really clean and pleasant taste. Note: You might think "I will add more sugar!" thinking brilliantly, no doubt, that you will get more alcohol. (which you do) The concern is that the taste deteriorates somewhat. The MALT is what gives it the flavor, the sugar does nothing but convert to alcohol.

If you are using the LAST recipe, that uses one Kilo of Sugar, I recommend adding more 50% more MALT and HALVING the Sugar to 500g. The point being, you can experiment, but always remember, the malt creates alcohol and flavor, sugar only creates alcohol. If you are extremely price conscious do not be misery on the malt with the notion you can just add more sugar to get a decent beer with higher alcohol. All you get is more alcohol and less taste.

The other cost not yet mentioned is the capping. Crown seals, as the beer caps are generally called, are not particularly expensive, and if you are carefully removing them from your finished product, they are reusable. Again, available on Ebay, and I also stress, careful removal of caps is a good idea at all times, just in case you made the novice mistake of too much sugar and subsequently get too much carbon dioxide in the bottles. I have done this, and have had the great joy of hearing "pop" "pop" "pop" as beer after beer exploded in my shed.

As you read through, you will discover how easy it is to recover yeast from the Fermenter. This you can use again and again. It is a small cost saving, but also it seems to improve the flavor.

"Paint Bucket" Fermenters. Be sensible though, add a TAP. It makes a world of difference. Note the use of TEMPERATURE STRIPS on the side.

"Easy Way" or BEER PACK?

The first decision most start-up brewers will need to make is whether they go the aforementioned suggestion of the "Easy Way" technique, or they spend more and go for a commercial BEER PACK. The first thing to say is that the majority of the commercially available kits are very good. You will not go wrong, but if it is economy and the ability to TAILOR your brew to suit yourself, then the "Easy Way" has to be the preferred option.

You will certainly not find the quality inferior, it is just a little more work. But having said this, once you have done the "Easy Way" a few times there is nothing much in it to challenge you, only it takes a little more time. Once you have boiled the MALT EXTRACT, SUGAR and HOPS for about thirty minutes, and strained it, the labor with the "Easy Way" versus a BEER PACK is the same.

If you are going to use the "Easy Way" recipe, it is time to decide what sort of HOPS you will use. The HOP EXTRACT is the easiest, but most expensive. The MALT EXTRACT and the SUGAR are brought to the boil and tipped into the Fermenter, then the HOP EXTRACT is added to the water. HOP PELLETS are a cheaper option, even cheaper than HOP FLOWERS, but they can be harder to obtain. (Editor Note: Not anymore. One Kilo for $50 on Ebay now)

The bulk of HOP PELLETS does not increase with boiling and there is little reason for straining. They must be boiled with the MALT EXTRACT and SUGAR for at least twenty minutes. HOP FLOWERS, on the other hand, expand with the boiling and must go for thirty minutes. You must also strain the result, no big problem. *The BIGGEST issue is where the hops come from.* Different countries have different strengths, plus this can vary from season to season. Here you would do well to ask your supplier for their advice. It is a thing only practice and experience can properly teach.

Choosing random suppliers for HOPS (of any variety) will introduce the most random element to the outcome of your brew. Cheaper is not necessarily worse, but as a rule, I avoid all Chinese imports. For no reason other than the pesticides they use are a significant risk factor, and there is little governmental oversight such as you will find with Tasmanian or local hops.

Your taste buds will be the determining factor in what you eventually settle on. (Editors Note: Summer Hops on Ebay supply, at a reasonable cost, AUSTRALIAN HOPS. They also give the Beta and Alpha Acid contents.) Whatever way you go, you will find it is best to settle on a supplier and stick with them if you wish to attain a consistent quality of homebrew.

Overall, BEER PACKS are simply more expensive, but more convenient. You may have already purchased some beer making equipment, and it often comes with a commercial pack. I recommend that you use it to get your feet wet, then try one of the "Easy Way" recipes, and do a taste comparison.

In the end, your taste buds tell you what works for you, not this book.

EQUIPMENT: What you need

There is no use going to war with a broomstick. You need to have the correct equipment and here I list everything that is essential to your brewing endeavors. It is easily available from any brew shop.

A large SAUCEPAN with TWO HANDLES. It needs to have a capacity of seven to eight liters (2 gallons) or more. This equates to a pan that is 280mm across and 140mm deep (11" x 5 1/2") There is probably one in the kitchen. Aluminum, enamel or stainless steel is sufficient, cast iron is not advised due to slow heating and weight. Aluminum ones are readily available in Asian food stores.

A FERMENTER: This is a container of at least 22.5 liter capacity (5.9 US Gallons) with a TAP and an AIRLOCK. You need one with a fully sealable lid, as you have to make it airtight. Larger is fine, and there are brands that do a six-gallon capacity.

SODIUM METABISULPHITE POWDER used to mix with water for a STERILIZING SOLUTION. The ratio to use is between one to two heaped tablespoons to two liters of water. Be warned, if you are ASTHMATIC you should not use product this for sterilizing. ***Ordinary household bleach** will also do the job perfectly well, but remember to always RINSE all compounds out before adding a brew*. There is a chapter on bleach later in this book.

HYDROMETER: You need a measuring device, and the HYDROMETER is the main tool for knowing your alcohol content. It also tells you when fermentation is complete and it is safe to bottle your beer. Simple observation will also tell you when the fermentation cycle is done, by the lack of activity in the airlock on the Fermenter, unless you are brewing and bottling in very cold conditions.

THERMOMETER: You need to have a thermometer that can be immersed in liquids.

BOTTLE CAPPER. Whether this is the simple "hammer on" version or a more elaborate lever action one, you need to be able to seal your bottles for the final cycle where the beer becomes a drinkable beverage. The simple 'tap on' capper is perfectly acceptable AND CHEAP.

BOTTLE WASHING BRUSH. Wire framed, similar to what is used in washing baby bottles. You need to be able to physically brush your bottles to clean them.

THIRTY LARGE BEER BOTTLES. You need something to put the result of your efforts into. STUBBIES are not recommended unless they are the thick glass variety. You will easily find yourself breaking the thin wall old fashioned small bottles. Obviously, you also need the equivalent number of crown seals.

There are several beer making kits on the market. *(Editors Note: Ebay has a few starting from under $90 going through to $160)* But be aware that many will have a Hydrometer, but not a thermometer, and vice versa. You need ALL THE ABOVE if you are going to make your own beer at home.

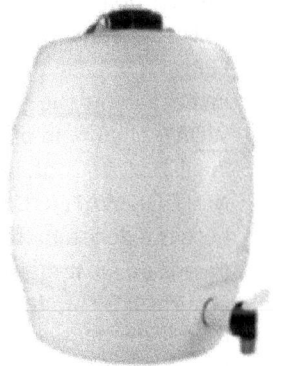

"Barrel" Fermenter

How this Equipment is Used

Fully detailed instructions will be given shortly, but first, it is important to know why you need the aforementioned items.

WATER is brought to boil in the SAUCEPAN on the stove. MALT EXTRACT and SUGAR (or dextrose/glucose) are added, stirred, and brought to the boil again. If using HOP EXTRACT for hopping, remove the saucepan/boiler from the stove then tip contents into the FERMENTER. The HOP EXTRACT is then added as you mix water with the starter brew in the FERMENTER.

If using HOP FLOWERS or HOP PELLETS bring MALT EXTRACT, SUGAR and water to boil as mentioned above, then add the hop flowers or hop pellets. This is SIMMERED gently for about thirty minutes with occasional stirring. NOTE: Fast boiling of MALT EXTRACT causes it to boil over.

Hop Flowers swell enormously when boiled and when boiling is complete they must be strained out and thrown away. A colander is sufficient for this. The flowers are discarded and the liquid goes straight into the FERMENTER.

HOP PELLETS do not have a residue and often appears to not need straining, however, all your initial brew mixes should be run through a fine mesh strainer when pouring into the FERMENTER. An ideal strainer is women's pantyhose. You can stretch it over an ordinary wire strainer. Be prepared for jokes from friends as to why you are purchasing pantyhose.

Remember, the FERMENTER must be cleaned, sterilized and rinsed before you add anything to it!

If you are using a commercial BEER PACK (can of concentrate) all you need do is bring 2 liters (1/2 gallon) of water to the boil, add the concentrate, and stir until dissolved. Remember, SIMMER boil only. Malt Extract will cause a fast boil to boil over. It is often recommended on the can to add the concentrate to water you have just boiled and decanted into the FERMENTER. This is not ideal, a gentle simmer boil while dissolving the concentrate always gives the best results.

Let's assume that whatever method you have used, the liquid is in the FERMENTER. Fill the 22.5 liters (5.9 US Gallons) FERMENTER to about 10 cm (4 inches) below the rim with water. This mixture has a name and is now called the WORT. The origin of this word is from Middle English and means 'root'. (Old English *wyrt:* root, herb, plant) In other words, it is where everything grows from. NOTE ON WATER: Obviously, filtered tap water is sufficient in most countries, but here is a decision. If you are concerned and wish to use spring water, etc. by all means do so. It makes little difference, but in Western Countries, tap water ensures it is safe for drinking. If water contamination is a possibility, you will need to boil and allow it to cool, using a few drops of FOOD GRADE hydrogen Peroxide to sterilize the water

Keeping the WORT at an even temperature, between the range of twenty-two to thirty-five degrees Celsius (seventy to ninety degree Fahrenheit), is essential. Here is where you use the STERILIZED Thermometer to keep track of the temperature. You can also use your STERILIZED Hydrometer to get readings on potential alcohol levels. (There is a chapter on Hydrometers)

You now open your packet of YEAST and sprinkle all the contents into the WORT. Stir in gently, then screw down tightly the lid of the FERMENTER and add water to HALF FILL the airlock. In twenty-four hours or less, depending on temperature, you will see gas bubbles popping through the AIRLOCK, which means the fermentation process has begun. All those ingredients you have used are now being turned into BEER.

When the bubbles are popping every two to three seconds, the fermentation process is in full swing. After a couple of days, the process will slow to one bubble every eight seconds or so, and now you will need to add some water to TOP UP the FERMENTER. Remove the lid, quickly add room temperature water to almost completely fill the FERMENTER and replace the lid IMMEDIATELY, in order to prevent wild yeasts or bacteria (germs) from gaining entry into your creation.

Most brewers I know DO NOT top up the FERMENTER with water but add instead around 150 ml (5 fl oz) of water to each large bottle when bottling. This method, although a little more trouble, has the advantage that the FERMENTER IS NOT OPENED AT ALL between starting and bottling. The beasties I have mentioned just love to get into your precious WORT and, obviously, by not opening the lid they are kept out.

If cleanliness and sterilization have been properly adhered to, it can be said that removing the lid of the FERMENTER is the MOST LIKELY CAUSE OF TROUBLE AND FAILURE in home brewing. As a note, Home Brew beer should always have a fresh, clean taste. Any 'off' or unpleasant taste is NOT NORMAL and is due to neglect somewhere along the line. Do not take shortcuts, it will result in a 'bad batch'.

In reasonably warm conditions, fermentation is usually done in four days. In cold conditions, this can slow the process down by a few days. Presuming your AIRLOCK has been bubbling vigorously, the fermentation process is complete when bubbles are 'popping' in the AIRLOCK at five minutes or longer intervals. At this point, the WORT is cloudy, so allow another five days for the fine particulate in the WORT to settle to the bottom of the FERMENTER, otherwise, you will find an unnecessary amount of sediment settling in the bottle. Usually, nine days after starting your WORT, the beer is ready to bottle.

A teaspoon of SUGAR is added to each large bottle before it is filled. The purpose of this is to start of a SECONDARY FERMENTATION PROCESS in the bottle itself. This is what produces the foaming head and the gas bubbles in your beer. As mentioned, if you choose to NOT add water to top up the FERMENTER, you would add 150 ml (5 oz) of water to each bottle with the teaspoon of sugar.

SUMMARY: In practice, boiling the water, simmering the MALT EXTRACT, adding the various versions of HOPS takes only forty minutes. Using a BEER PACK, it takes only five minutes. Bottling takes around thirty minutes. Not a lot of time or money is required for you to get thirty bottles of good, cheap beer.

DECIDING WHAT TO BREW

I t is now time to ask yourself what type of beer you want to brew. This is whether to use a COMMERCIAL BEER PACK or one of the "EASY WAY" recipes. If the later, you need to choose the type of HOPS you will use - HOP EXTRACT, HOP PELLETS, or HOP FLOWERS. We will be giving you STEP BY STEP directions on what to do, and all you need worry about is making sure you 'follow the bouncing ball', so to speak. Please note, even when using a BEER PACK, the directions still apply.

Right now, all you need do is make sure you have all the necessary ingredients and the correct equipment. But there is something you should check before all else, and this is the condition of your FERMENTER. You must be sure the seal is good because a poor seal will ruin your beer.

To do this, fill it with tap water to around 10cm (4 inches) below the rim. Screw the lid on, fill the AIRLOCK half-full with water - everything you would normally do when making your WORT. Now, with the AIRLOCK in place, apply gentle pressure with your fingers to the top of the lid and maintain this pressure for a few seconds. The water level in the AIRLOCK should rise up and stay at the same level when you apply pressure to the top of the lid.

If it does, the FERMENTER is sealing properly. If the water level in the AIRLOCK drops quickly while you are adding pressure, you have an air leak. Check the lid is screwed on properly (You can easily cross thread the plastic ones) and that the RUBBER SEAL is flush with the socket. Next check that the RUBBER GROMMET on the AIRLOCK itself is good. (sometimes it needs Vaseline to help seal). If it all checks out, go through the procedure again. Remember, all you need do is apply enough pressure to cause the water in the AIRLOCK to rise, and keep the pressure on to see if it drops. Once this works, you know the FERMENTER will seal properly.

Now STERILIZE and RINSE the FERMENTER. Sterilizing is dealt with in the following chapter. When this is done, immediately add some water, to the FERMENTER and the AIRLOCK, and seal it up, to be ready for use. Don't forget to sterilize the lid, the rubber seals, and the AIRLOCK, and also run some solution through the tap itself.

IT CANNOT BE EMPHASISED ENOUGH how wild yeasts and bacteria can, and will, ruin your beer if they get a chance to get in.

***Adding MALT EXTRACT
to the WORT***

STERILIZING

Everything about Home Brewed Beer pivots on how CLEAN your equipment and process is. It cannot be stressed enough. Also, the basic method of sterilization uses SODIUM METABISULPHITE which can cause ASTHMATICS a concern. If you are asthmatic, use household bleach. (There is a chapter called "Household Bleach - The Brewers Friend" later in this book) All utensils that are introduced to the WORT at any stage must be thoroughly cleaned and sterilized. If they are not, you will get 'bad batches' and never know what the problem that created it might be.

The 'problem' is usually laziness, carelessness, and inattention to detail which allows bacteria and wild yeasts to get into your WORT. The first step is ensuring that they are not there in the first place! In this section we are going over the use of SODIUM METABISULPHITE, a product usually incorporated into most beer making kits. We also cover this in the section on "Bottle Washing" but the point on sterilization bears repeating again and again. It is core to the brewing process.

Use two tablespoons of SODIUM METABISULPHITE powder per two liters of water for a strong solution. If you want to use this for rinsing, use one tablespoon per two liters (1/2 gallon) of water. Most will use tap water for rinsing, or a weak bleach and water solution. The instructions for mixing are generally on the packet.

To sterilize the FERMENTER, fill the AIRLOCK with some of the solutions, and tip some into the FERMENTER. Then take the lid, and pour the solution into this as well. (Don't forget the lid!) Let some of the solution in the FERMENTER run through the tap. Wash the seals in the solution, then screw it all together, and shake the solution thoroughly inside the FERMENTER. Make sure all surfaces are washed inside the FERMENTER. Using the tap, drain the fluid into a container to hold it.

Now take it apart and wash every part of the FERMENTER, including the seals and AIRLOCK. I often use a garden hose for this process, quick and simple. You don't need to dry anything, just drain and screw the FERMENTER back together, making sure to add water to the AIRLOCK. The FERMENTER is now ready to use.

The SODIUM METABISULPHITE solution can be used many times, so storing it is a good idea. Any container works. There is a strong smell it gives, and when that fades, the solution is too weak to reuse. Take care when smelling it, as breathing deeply can and will irritate the nose and throat. Again, do not use if Asthmatic.

When sterilizing bottles, it is a good idea to use a teacup strainer when decanting the sterilizing fluids, to catch any solids. Thermometers, Hydrometers, hydrometric jars, CROWN SEALS (most people forget the seals) and any and every utensil that will come in contact with your WORT must be sterilized and rinsed.

NOTE: There is a vast difference between something being CLEAN and something that is STERILIZED. Something that is cleaned will still have bacteria on it. Sterilizing means that bacteria (germs) have been killed. It doesn't take a whole lot of effort to do so. It will make an ENORMOUS difference to the quality of your brew when you do, however.

NEGLECT **WILL** RESULT IN POOR QUALITY, OR EVEN UNDRINKABLE, BEER!

BREWING - Easy STEP by STEP Instructions

*F*rom here on, everything will be stated in the fullest detail, so that, hopefully, no questions will remain unanswered. You will not be left asking yourself "What does he mean by that?". This book is written with the person in mind who has never made beer before and who is not able to understand complicated incomplete instructions.

I heartily recommend to you to keep a BREWERS NOTEBOOK, and write every single detail, no matter how small or insignificant, at every step of the way. It will be an INVALUABLE tool for analyzing any problems if any, and it will be helpful in keeping a clear record of any variations you try. A simple notebook will because an incredible tool for your memory as the years roll past.

Follow these steps to make a perfect brew

STEP ONE: Bringing the Water to Boil on the Stove

Despite what most BEER PACKS (cans of concentrate) will say, it is always best to bring water to boil on a stove and allow it to simmer when mixing ingredients. Take two liters (1/2 gallon) or water - the exact quantity is not important - and bring it to boil. In an electric jug, on the stove, it doesn't matter. Decant into your SAUCEPAN in preparation to add the ingredients. The only thing that matters is you have approximately two liters of water simmering away.

STEP TWO: Adding the Ingredients

If using a BEER PACK, dissolve the contents into the simmering water. Add SUGAR or DEXTROSE is you are using it, remove the boiler from the stove, and stir it in. Proceed to STEP THREE. (Note: It is likely that the beer from your BEER PACK will have a slightly higher alcohol content) When using a BEER PACK, there is no need to bring the water back to the boil. Incidentally, there is no need to STERILIZE the boiler, as the boiling water does this.

When using the "Easy Way" your choice of HOPS determine the procedure. Add your MALT EXTRACT to the simmering water, and stir in till it dissolves.(Hint: placing your spoon you will use to draw out the Malt Extract into the hot water first will make it easier to scoop up the Malt Extract) Then ADD SUGAR or DEXTROSE as per the recipe you choose. Stir in till dissolved.

HOP FLOWERS, or HOP PELLETS, are then stirred into the simmering mix. Bring the water back to the boil as you stir it in, then simmer the whole mix gently for thirty minutes.

IMPORTANT NOTE: Boiling MALT EXTRACT on high heat will cause it to 'broil' and overflow the saucepan. It will also BURN on the bottom of the pan and your brew is effectively compromised. Once the ingredients are mixed, SIMMER the liquid to keep it at heat.

STEP THREE: Creating the WORT

DECANTING the Starter Liquid. First, take care, it is hot. You must strain it twice, first with a colander or steel mesh strainer, and once more through pantyhose or similar FINE mesh. It is best to do this into a holding container and fine strain from there into the FERMENTER.

Whether you are using a BEER PACK or HOP PELLETS/HOP FLOWERS the process is the same, simply pour the contents into the FERMENTER. Obviously, the FERMENTER has to be of sufficient strength to cope with boiling water. I have heard of people using open rubbish bins for Fermenters, which is completely stupid.

If you are using HOP EXTRACT, I recommend that you mix this into the water you will be pouring into the FERMENTER to top it up. You can mix it in the Fermenter, the only thing that matters is that it is thoroughly dissolved.

Where conditions are warm to cool, all you have to do now is add the cold water to top up the existing WORT in the FERMENTER. Again, in a 22.5 liters (5.9 US Gallons) FERMENTER, you fill till you are 10cm (Four Inches) below the rim.

Using your sterilized THERMOMETER, the temperature of the FERMENTER CONTENTS (now called 'The WORT') should be within the range of twenty-one to thirty-five degrees Centigrade. (seventy to ninety-five degree Fahrenheit) If so, proceed to STEP FOUR. If not, you must allow it to cool to the appropriate temperature, but remember, you must SEAL the FERMENTER at all times to prevent wild yeasts and bacteria from getting in.

TEMPERATURE IS IMPORTANT! Now, we presume the reader may be living anywhere in the world, and potentially be dealing with freezing cold to extremely hot temperatures. In a very hot situation, before adding the YEAST you may need to place the FERMENTER into the fridge to cool it, however, I have made a WORT where temperatures were forty-five point five degrees Centigrade. (114 degrees Fahrenheit) This is detailed in the chapter entitled "Yeast".

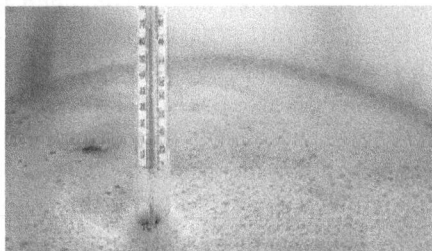

Whatever way you work this, either by heating the bulk of the water to an appropriate temperature or cooling it before adding to the FERMENTER, you need to aim for the aforementioned temperature range before proceeding to add the YEAST. You also need to KEEP the Wort at this temperature, so it may mean leaving it in an air-conditioned room, as one possibility.

In very cold situations it is somewhat easier. A simple solution to get the correct starting temperature is to boil more water in the pre-mix. Whatever way you do this, you use your THERMOMETER to ensure you are in the correct temperature range. Also, aim for the MAXIMUM starting temperature of thirty-five degrees C (95F) assuming that the WORT int he FERMENTER will cool more rapidly.

Keep the FERMENTER in a room with constant temperature, or in cold situations, wrap it in an electric blanket. Another way in cold situations it to keep the FERMENTER in a box with an old-fashioned incandescent bulb inside it to keep up a heat supply. Rules of common sense apply, do not allow a hot light bulb to come in contact with a plastic FERMENTER or a cardboard box.

If your WORT cools down too much, it will not 'kill' the beer, but it will slow down the fermentation process significantly. It may even put it in park, and will not start again until a suitable temperature has been reached. Otherwise, at the other end, a WORT seems to be able to handle temperatures of up to 45.5C (114F)

STEP FOUR: Pitching (adding) the Yeast

Take your packet of YEAST GRANULES (Brewers Yeast) and sprinkle the contents on top of the WORT in the FERMENTER. There is no need to stir but do so if you wish. With your next brew, if you wish to save money, you can recover this yeast from the sediment and re-use it. This is a different procedure discussed later in this book. Essentially, yeast is a LIVING ORGANISM. It GROWS in the process of fermenting your beer. You can use the YEAST you just sprinkled in over and over.

This is when you screw the lid of your FERMENTER down tight. Fill the AIRLOCK half-full with water, and I recommend adding a small amount of SODIUM METABISULPHITE SOLUTION to this water. It prevents bacteria getting in through this only open hole to the WORT.

Now check that the FERMENTER is properly sealed, using the pressure on the lid technique described in the previous chapter. Keep a gentle pressure on the lid, and watch to make sure the water in the AIRLOCK rises and stays at the same level as the pressure you apply. Once you have ensured that the FERMENTER is not leaking air you can proceed on to STEP FIVE.

If not, there is a problem that must be addressed. It is most likely a SEAL that has not sat properly. You need to solve this quickly, so a brief inspection of seals and the application of Vaseline, or similar, may solve the problem. You really want to be certain that your FERMENTER is not leaking air.

If there are any problems at this stage, decant the WORT into a sealed container and go through the entire process of getting your FERMENTER sorted. If you can't get it sealed properly you may as well go on to Step Five and just hope for the best.

Step FIVE: Fermentation

Fermentation is the process by which the YEAST converts the MALT and the SUGAR in your WORT into alcohol, thus transforms your WORT into BEER. Presuming you have done everything correctly up to this point, TEMPERATURE is the main factor during fermentation. If the WORT TEMPERATURE (not air temp) gets TOO HOT, it can kill off or degrade the YEAST. If it gets TOO COLD, the fermentation process itself will simply stop. An ideal temperature is around twenty-one degrees Centigrade (seventy degrees Fahrenheit)

If you have the FERMENTER indoors, this naturally tends to regulate temperature. If you have it out in a garden shed, as one example, you will need to pay attention to this factor. As discussed, in cold weather you can do things like wrap an electric blanket around the FERMENTER, in hot weather perhaps consider an air-conditioned room.

Inside of twenty-four hours after initial sealing of the FERMENTER, the process should have started. Gas bubbles will form in the AIRLOCK, and they will 'pop' slowly at first, building to a regular bubble through the AIRLOCK every one to three seconds when the fermentation is well underway. This is carbon dioxide escaping and it is a bi-product of the sugar converting to alcohol. The bubbles will be occasional at first, building up to a 'blip' every one to three seconds, then settling down to an occasional bubble every eight seconds until fermentation is complete, usually inside nine to ten days after the FERMENTER is sealed.

The entire process is one where the YEAST is acting upon the SUGARS in the WORT. As the SUGAR is converted, the fermentation cycle finishes. This YEAST getting to work is the heart and soul of the entire brewing process. We have a trouble shooting section, but we will cover the basics here.

If the bubbling has not started in the AIRLOCK inside twenty-four hours after adding the YEAST (Assuming the WORT TEMPERATURE is at the correct level) you need to take the lid off the FERMENTER. If the surface of the WORT is covered with a brown, bubbly froth, or if you can see bubble activity, things are proceeding as normal. It will be a lack of a proper seal in the FERMENTER.

The WORT will probably still be fine, but refit the lid, check seals, etc. and DO NOT open again until fermentation is done, which is bottling time. Do NOT top up the FERMENTER with water, but add some water to your bottles with the sugar when bottling them. You will usually find that when you replace the lid, the bubbles will start.

If, on removing the lid, you find the WORT does NOT have froth or bubbles and is flat and clear, then there is NO ACTIVITY. Fermentation has NOT begun. The WORT may be too cold or too hot, or you may have put in the wrong type of YEAST. Temperature you can correct, if it is the wrong yeast, you need to reseed the WORT with a different brand and see how it goes.

Healthy Fermentation - second day

However, if you have followed instructions, made everything sterile, and performed things in the order they must be done, it is extremely unlikely you will experience troubles like this. Home brewing should be an enjoyable, fuss-free hobby which provides endless satisfaction and enjoyment.

Step SIX: Topping up the Fermenter

When the first furious level of fermentation is done, when the AIRLOCK is popping every ten seconds or so (about the 3rd or 4th day usually) I recommend you have ready five to six liters (one gallon) of water for TOPPING UP THE FERMENTER. Note: In very cold conditions, this should be warm water.

You want everything ready because the ideal is to have the top off the FERMENTER for the SHORTEST TIME POSSIBLE. This is to reduce the risk of infecting the WORT with wild yeasts and/or bacteria. They are there in the air, just hungering to get in and breed in your WORT.

However, while you are there, it is a good time to check on proceedings. There will be a froth, or foam, on the top of the WORT. If there are some clear patches, you will see fine bubbles rising to the surface. This is a clear sign that fermentation is underway. While this is happening it is ABSOLUTELY not time to bottle. It is a very immature beer, at this stage.

Take your water and fill the FERMENTER to within three cm (one inch) of the rim, then screw the lid back down tightly. This is called TOPPING UP and it is the ONLY TIME you would normally remove the lid of your FERMENTER. The evolvement of CO_2 gas along with the froth on the top of the WORT protects it from contamination to a degree at this stage of the process, but apart from lack of sterilization and cleanliness, taking the lid off your FERMENTER is the main source of problems.

On this point, if your FERMENTER is out of doors, open to the wind and elements, it is better to avoid this topping up process altogether. add 150 ml (5 fl oz) to each bottle prior to bottling instead. (assuming large 750 ml - 26 fl oz - bottles are being used)

Another reason to avoid the topping up process is that you have forgotten to do it at the correct time, or you have been away. If you open the FERMENTER and see the process is not working properly, again, do not top up. If the fermentation process is complete, one bubble every ten seconds or so, then DO NOT TOP UP.

Many brewers do not top up at all, preferring to add water to the bottles at bottling time. It is a little more trouble, but it does avoid the risk of contamination. Personally, I like to inspect the brew at the crucial turning point just before full fermentation.

You may well ask, and rightly so, 'Why not fill the Fermenter right up PRIOR to Pitching the Yeast?' This is because some air-space is needed at the top of the WORT to accommodate the big head of froth produced with the first stage of vigorous fermentation. If your FERMENTER is filled too close to the lid, the froth will come pouring out the AIRLOCK. (More details in the Trouble Shooting section)

Of course, ALL of this would be avoided if the MANUFACTURERS of FERMENTERS would consider the needs of their clients and simply make them 27 liters instead of 22.5 (six gallons) which would mean we could all brew without the need to either top up or add water to bottles. *(Ed Note: Ten Gallon now available)*

Regardless, topping up is generally perfectly safe as long as it is done while fermentation is still active, and if it is done quickly in a controlled environment.

Step SEVEN: Allowing the Fermentation to Finish

Rather than repeat everything for cold or hot ambient temperatures, and assuming the average, within target temperatures are maintained leave the WORT for a further seven or eight days after Topping Up. After four days (Longer if it is cold) the fermentation cycle will be complete, and the extra time is to allow sediment to fall to the bottom of the FERMENTER. It is far better there than in your bottles and this extra time results in a much cleaner and clearer brew. Most 'cloudy' home brews occur because someone bottled too soon after fermentation was complete.

Step EIGHT: Bottling

Presuming you have selected the type of bottle you wish to use, the final stage is bottling your brew. I recommend the standard 750 ml (26 fl oz) beer bottle, and your standard brew will need thirty of these. Obviously, if you are using stubbies, you need sixty bottles. Make sure your bottles are scrupulously clean and sterilized. Also, have a marker pen for noting on the bottle cap the date you are bottling your brew. Have ready your bottle capper, a hammer and a thick piece of wood to support the bottle while hammering the seal on. (NEVER hammer a seal on with a bottle resting on concrete. It will break.

Prior to bottling, remove the lid from your FERMENTER to check that the process is complete. Again, presuming the brew was active for three to four days and bubbling away, then it settled down and stopped, and you have allowed a few days for the sediment to settle, everything will be ready to proceed. The issue is usually that a spell of extremely cold weather that may have stopped the fermentation.

With the lid, off, look down into the fermented beer (It is no longer called the WORT) and if the surface is clear, with perhaps only a few 'islands' of bubbles floating there, you are good to go. You should be able to see into the liquid for a short distance, in other words, it is reasonably clear, not cloudy. This means fermentation is done, and sediment has largely settled.

For the inexperienced, take your sterilized Hydrometer and float it on the surface. It should read 1.003 to 1.006. Commercial BEER PACKS tend toward the higher number. If for any reason, the Hydrometer reading is above 1.01 or higher, bottling should be put off for a few days. *NOTE: Bottling your beer with a Hydrometer reading of higher than 1.080 will invite bottles to explode.*

If you have not TOPPED UP, you may or may not wish to add water to the bottles. Presuming you haven't topped up, you will need some funnels and a measuring glass. You need to add 150 ml (5 fl oz) to every bottle. It is NOT essential to do this, but the math is fairly simple, 150 ml x 30 = 4500 ml or another six bottles. One funnel is for adding water, the other is a dry one for adding the sugar.

The SUGAR you add at this point is what creates the BUBBLES in your beer, and it helps with the alcohol conversion. A teaspoon is all you need as a measuring device. Add a FLAT teaspoon to every bottle (5 ml / 0.2 fl oz)

A small hint: If you use a hacksaw and cut off your funnel at the point where it is the correct size for your bottles, you give yourself the widest possible opening for the sugar to flow through. The funnel you use for liquids does not require this, but you are well advised to stretch a rubber band around the stem, just below where the funnel would normally sit. The trick is to OFFSET the rubber band as this provides a gap for the displaced air from the bottle to leave. It makes filling the bottle much easier and quicker.

PRIMING THE BOTTLES: This is the term used for adding the SUGAR, to set up the secondary fermentation process INSIDE the bottle. There is a little carryover yeast from the initial process still resident, and this will react with the SUGAR you just added. The secondary fermentation is what give the foaming head, the fine bubbles and the 'life' to the beer when you open it up to drink.

Priming should receive your full care and attention, as it is where you achieve the final product that all your friends will want to drink. Too little sugar and you get a 'flat' beer. Too much sugar will cause the contents of the bottle to erupt when opening, or even cause them to burst in storage. This can cause a chain reaction and you can lose an entire batch, plus pose a significant health hazard from shattered glass. A worst case scenario is the bottle bursting in your hands as you prepare to open it. All this is avoided with a little care.

The teaspoon is the simple measure, and for those who have no clue as to what this is, it is the smallest average spoon in the cutlery drawer. It is technically 5 ml or 0.2 fl oz. The problem is, there is no 'set' teaspoon, and you see them come in a variety of sizes nowadays. Find a 5ml measure, take a number of small spoons from your drawer, and find the one that suits by pouring fluid from the measure into each spoon. Keep this spoon and set it aside for all your brewing needs, and we will never have to have this conversation again!

Now, given all this, your personal taste as to how much PRIMING SUGAR to add to the bottle will change from person to person. Start with a ROUNDED teaspoon, and adjust to taste on different batches. If the beer is flat, you didn't add enough. (that would have been a LEVEL teaspoon) If it is foaming too much when opening, you added too much. (this would have been a HEAPED teaspoon) You cannot change the finished product once you have opened your bottle for drinking, but remember, it is CRUCIAL to the final finish of your beer to get this PRIMING SUGAR addition in your final bottling correct.

Another small note: If you are topping up your bottles with water, PUT THE PRIMING SUGAR IN FIRST. Otherwise, the sugar sticks to the neck of the bottle. Obvious and simple, but important.

The type of SUGAR you use for PRIMING is not particularly important. You can use raw sugar, even honey, but you must have the measurement exact. Powdered DEXTROSE or powdered GLUCOSE may also be used. Dried Malt Extract is also available in the same quantity. (Liquid Malt and honey is difficult to gauge as to the correct quantity)

Honey makes an EXCELLENT Primer Sugar and must be used in the amount of TWO LEVEL TEASPOONS per bottle.

I recommend, for simplicity, that you stick to simple white sugar until you gain more experience.

Some brewers go to the length of sterilizing the sugar by boiling it, then adding the correct amount of this fluid to add to each bottle, but this is extremely fussy and very open to mistakes. Their thoroughness is to be commended, but I assure you, nothing much can live on white sugar and it is effectively already sanitized. It is pure carbohydrate.

However, if you really do want to add boiling water to your priming sugar, 225 grams, or 8oz, would be the quantity for 30 large bottles. It really is unnecessary, however.

NOW, we finally get to actually adding the SUGAR. Take your time! Do NOT rush this process, because it is very easy to forget where you are and accidentally add a double teaspoon of sugar, etc. Or worse, you miss a bottle entirely! It can easily happen if you allow yourself to become distracted. As a note, some people like to use the LITRE size plastic bottles. If so, you must add one third more sugar. Again, it is obvious, but we want to cover all bases.

If you have not TOPPED UP the Fermenter, add around 150 ml (5 fl oz) of water to each bottle. The exact amount is not so important, but overall it will give you six more bottles than if you do not do it. Taste is much the same with or without the added water. Now you are ready to start the actual bottling process. You have established that the fermentation process is complete, you have let everything settle for a few days, and now you must REPLACE the lid of your FERMENTER

First, have the FERMENTER sitting on a table at a height that works for you. You will need a stool or box to sit on. Now check that the tap is free of sediment, take a glass, bring it to the tap, and turn it on FULL briefly. There will be some sediment flushed out, with possibly some lumps of yeast. Turn the tap OFF, and reserve the tumbler, you may need the contents to top up the last bottle.

Using a CLEAR 30mm HOSE to make bottle filling easier

Next, take a 30 cm (12 inches) length of sterilized hose which you will have previously checked to make sure it fits over the nozzle of the tap on your FERMENTER, but inside the neck of your bottles. (ordinary garden hose is usually a good fit, but it is better to use a CLEAR hose) Connect the hose to the FERMENTER, and now you are ready to start filling your bottles. Editors Note: You can buy a length of hose with a pressure valse already fitted to it in a number of home brew beer shops. There are many options, including SYPHON hoses for under $10.

Take each bottle individually, bring it to the tap while inserting the hose into the bottle, and fill right to the brim. When you remove the hose, the level will drop to the correct point. (approx. 50 mm - 2 inches - from the rim) You can, of course, simply take the bottle to the tap and fill directly, but using the hose

means less contact with air, thus less oxidation. Plus using a length of hose speeds up the process, it is simply quicker and there is no froth to contend with.

Once you have filled your bottles to the 50 cm - 2 inch - below the rim mark, place each bottle you are about to cap onto a solid piece of wood. Place a crown seal over the top, and using the Hand Bottle Capper and a hammer, tap the seal down and into place. You will feel when it will go no further. Do NOT use excessive force, the seals can be 'tapped' on using largely the weight of the hammer. Check your seal on the first few by tipping the bottle sideways. You will soon find out if it is leaking. You MUST get a good seal, otherwise, the carbon dioxide of the final fermentation will escape, and you will have flat beer to drink.

Give each bottle you have sealed a few shakes, to help dissolve the SUGAR you added, and continue bottling until finished. You may need to tilt the FERMENTER carefully with the last two bottles, to get the last of the fluid out. Be careful, you don't want sediment going into your bottles. On this point, I recommend using CLEAR hose because you can see when the contents turn murky. Do not bottle heavily sedimented beer, because you may turn into the "Sedimental Bloke". That is the man who has their friends picking their teeth after drinking his beer.

MARK the bottle caps! Just the DATE is sufficient, and the last three you bottle, put an 'x' on them to know these bottles may have extra sediment. As a brewer who had over two thousand bottles under the flats I owned, not having every beer dated becomes a significant issue. You need to know how old your brew is, and if there is a bad batch (it can happen up to one in ten fermentations) you need to be able to identify every beer in that batch.

As a note, one of the reasons I got interested in adding a section on how to make COMMERCIAL BEER was because I was chatting with the man who ran Castlemaine Perkins beer. (Americans will love that we have a beer here called "XXXX") I asked his advice, saying that one in ten of my brews was sub-par. Not bad, just not as good as the others. He laughed, saying that was amazing. When I asked him why he said: "Why do you think we run ten vats per brew cycle?"

Apparently, they get the same, but they just mix it all together and it all comes out with the same taste in the end.

When bottling, using a clear hose, you will see a slight cloudiness in the beer. This is perfectly normal. When you see the clarity CHANGE, this is when you are getting to the sediment. The longer the FERMENTER sits after it has completed the fermentation cycle, the clearer the beer will be, but four days after the end of the cycle is more than sufficient. You, in fact, want a little bit of yeast still in suspension so that the secondary fermentation cycle in the bottle works properly.

Now, prior to bottling, when you were checking your beer, if you noticed a thick, white mold over the surface, or an oil-like slick with large bubbles and maybe some lines, go to the trouble shooting section before doing anything else. Otherwise, you are good to go. I will say that the first time I got a brew through to the bottling stage, I had a marvelous sense of satisfaction. It is not just about cost saving, it is also about achievement and knowing you have created something.

Step NINE: -Storing the Bottled Beer

The process of storing the bottles of beer upright to develop gas and mature is known as CONDITIONING or MATURATION. This is also called the Secondary Fermentation Cycle. It is a very similar process to your FIRST cycle, in the FERMENTER, when you added YEAST to the WORT. The WORT is basically a solution of SUGARS. Upon bottling, you added some SUGAR which will now react with the carry-over YEAST, still in suspension in the bottle. It produces a little more alcohol and CARBON DIOXIDE Gas, the thing that creates the 'bubbles' and the Foaming Head of your beer.

In the FERMENTER, this CO2 was able to escape through the AIRLOCK and was the cause of the bubbles that told you about the progress of your fermentation. Now the bottle is SEALED, so the gas CANNOT ESCAPE. It builds up pressure and infuses into the beer. It takes around four days under reasonably warm conditions to build up to full pressure.

It takes the gas several more days, up to three weeks or longer, to form in the beer. This is where the foaming head you like to see so much in your home brew is developed. During this period the contents in the bottle is losing the raw, not so pleasant taste it had when originally bottled. It slowly becomes perfectly clear, as the suspended yeasts and proteins sink to the bottom of the bottle, settling as it will into a reasonably firm matt. *NOTE: If there is a significant amount of sludge at the bottom of your bottle when you pour, change your brand of yeast.*

The purpose of standing your bottled beer upright is so that all the sediment will settle on the bottom and make pouring easier. It is NOT POSSIBLE to have home brew free of sediment unless you are able to filter it and then artificially carbonate it like commercial brewers do. Once you get used to pouring carefully, it is really not a concern.

One COMMERCIAL BREWER, Coopers of South Australia, produce their beers using the SAME METHOD as I have just described. They claim that the "Natural Sediment" is an added attraction, and I will say, they make an excellent beer. Incidentally, the live yeast in their beer is excellent for home brewing.

Your Home Brew will improve with age, up to a point. It will keep for a year without drama if left in a cool, dry spot, one not subject to direct light. However, there is no point in keeping it longer, it does not continue to improve.

In cold conditions, the stored bottles may need a warm place for a few days. (Some in cold countries store the freshly bottled beer in the hot water closet, or similar) After four days or so, the basic reaction with the yeasts are done, and temperature doesn't matter so much. Obviously, frozen is not good. Real Lager is often stored at near-freezing, but not immediately after fermentation. (The word "lager" comes from the German: *To Store*) Ideal storage temperature is around 10C (50F). Beer MUST NOT be stored where it will be in the sun, or it will quickly become undrinkable. *NOTE: One of the basic reasons beer is stored in colored glass bottles is to reduce the effect of light on the beer.*

Further, if you have bottled using CLEAR GLASS, it is imperative the bottles are stored in a dark place.

Step TEN: Recovering the Yeast

After you have bottled your beer, you will have all the yeast you will ever need. The small amount of YEAST that you added to the FERMENTER has now GROWN, be feeding on the SUGARS you gave it. Please note the Chapter on "Yeast and Re-Using Yeast, but in essence, this is a living organism that you can cultivate and re-use again and again. Some say the secondary yeasts produce a better beer, but I suspect this is more liely that the brewer has become more competent with practice.

Healthy "Yeast Cake" you can recover and reuse

The resultant sediments at the bottom of your FERMENTER usually take on one of two characteristics. The most common is for the dregs to be like a thick slurry which, if swished around with the last remaining beer, can be poured out of the Fermenter. Simply pour into a STERILIZED jar. With this type of dregs, it is difficult to distinguish between pure yeast, which has a creamy color, from the predominantly brown dregs with which it is mixed.

The other type of dregs are those that have settled to a firm MATT at the bottom of your FERMENTER. This cannot be poured out. If you gently disturb the sediment with your fingers you will see the YEAST as a cream, white or pale grey colored paste. Wash your hands to reduce the risk of contamination, and scoop out this yeast and put into a sterilized jar.

If you intend to keep either for an extended time, put it in the freezer, but be sure to leave an expansion gap to allow for it to freeze. Frozen Yeast will keep indefinitely. If you are going to reuse it within a few weeks, top up the jar with water, to stop contamination, and leave in the fridge.

When you next set up a brew, take two or three good dessert spoons of the dregs and pitch in (add) to your WORT. If the temperature is extremely cold, you can use the lot to get your brew off and running. The only reason for using a tiny amount of YEAST, in the beginning, is purely for economy.

If you have recycled the 'paste' type of yeast you can, with careful washing, separate off the brown and spent yeast cells from the living light colored ones. That said, if a WORT has become contaminated with wild yeasts or bacteria it will show evidence of this on the surface. There may be an oil-like film on the surface and large bubbles, or it may be covered in a white mold-like substance. In this instance DO NOT keep or reuse the yeast.

If you have been fortunate enough to get a good YEAST, with careful handling it will reproduce faithfully for years. Some are better than others and there is no rule of thumb to follow in selecting the best, only that it works well in your WORT.

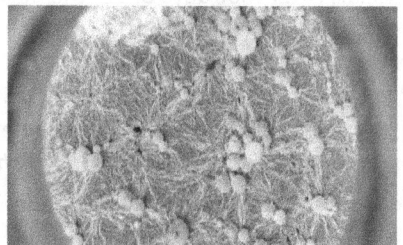

Contaminated Yeast Example

Step ELEVEN - Cleaning the FERMENTER

I always refer to this stage as the "Eleventh Step to Happiness"

Take the empty Fermenter out to the garden, and hose it out. There is nothing harmful in the dregs that can harm the grass. Now take a NON-METALLIC scourer (plastic) and use this to thoroughly clean the inside of the Fermenter. Keep this, and use it solely for this purpose. Give it a good scrubbing, and make sure you rinse it clean. Run water through the open tap and make sure it is free of sediment. Take out the rubber sealing ring in the lid and wash it as well. Flush the AIRLOCK and scrub the rubber seal it uses as well.

Take your bottle of SODIUM METABISUPHITE SOLUTION that you have saved from your sterilization process at the start of this journey and pour some onto the sealing ring and the lid, as well as the AIRLOCK. Then tip the rest into the Fermenter. Give it a good swill around and let some run through the tap. Return the sterilizing solution to its container, ideally, use a funnel. Try not to get too much in your lungs, it can irritate the membranes. Remember, if you are an ASTHMATIC do not use this, use household bleach instead.

If you inhale the SODIUM METABISUPHITE SOLUTION it gives you a belt up the nose similar to inhaling ammonia. So avoid this! If this strong smell has faded, the solution is spent.

Now close the tap and rinse the Fermenter and the lid well with water from the hose.. Replace the rubber ring, screw down the top, put some water in the AIRLOCK and put it into place so that it is ready to be used again. If it is going to be some time before you reuse the equipment, put a little of your SODIUM METABISULPHITE SOLUTION in the unit before sealing. (Not bleach, it can break down the plastic)

That is IT! You have done your first brew from start to finish. I am sure you will agree there is little that is difficult in the entire process only, perhaps, fear of the unknown. Now it is a known, and you will be good to go for your next lot. You can now tackle your next brew with confidence and enthusiasm.

SUMMARY: Boiling up and preparing the WORT, using either the "Easy Way" or a BEER PACK really takes a maximum of forty minutes. Topping Up takes a few minutes. Bottling takes thirty minutes. *All up, for under an hour and a half of labor and a small investment, you will have two and a half cartons of beer.* It is not a great investment of time, even for the busiest of people.

If you like beer, what a pleasure it is to admire it in the glass, knowing that this is something you alone have created. If you have done it right, you have the enormous bonus that it is good to drink as well!

Chin Chin!

Beer, it's the best damn drink in the word!

Jack Nicholson

TROUBLE SHOOTING SECTION

The various troubles that can arise may be divided broadly into two parts - those which occur PRIOR to bottling, and those which occur AFTER bottling. Let us deal with the problem in the order they are likely to arise.

PROBLEMS BEFORE BOTTLING

Scenario One: *You have taken the WORT from the stove and are tipping it into the FERMENTER.* The WORT is running all over the kitchen floor. Logically, dumkoft, you have left the TAP open on the FERMENTER. This usually occurs only once in your brewing career. A second occurrence means a change of meds is called for.

Scenario Two: *There are NO GAS BUBBLES in the AIRLOCK within 48 hours of adding YEAST.* First, did you put water in the AIRLOCK? Second, did you put SUFFICIENT water in the AIRLOCK? If so, the most likely cause is that the FERMENTER is not sealed properly. (hence the suggestion that you test it BEFORE adding the WORT, and TESTING the FERMENTER SEAL after you add it)

Response: Remove the lid and look inside. If there is a thickish brown scum on the top of the WORT, the YEAST is working OK. The problem is that the CO2 gas is leaking elsewhere. Replace the lid carefully, and screw it down properly. Unless you have a crack in the lid, or a broken seal, there is no other place for the expanding gas to go but out through the AIRLOCK. Chances are no great hard has been done, and fermentation will pick up in the coming day. In such an instance, OMIT THE TOPPING UP PROCESS. Just add some water to your bottles when bottling.

If there is NO brown scum on the surface of the WORT, the YEAST is not working. This may be due to very low temperatures in the WORT when adding the YEAST. Raise the temperature of the WORT, by leaving it in the SUN or wrapping in an electric blanket, and the YEAST should start working. If it doesn't, try a different type of brewers yeast.

Yeast will survive freezing, but it will not survive high temperatures. The other possibility is that you added the yeast to the boiled water of the WORT directly, and killed it. Alternatively, if you are in an extremely hot environment, it may have killed off the yeast. Again, add more YEAST and move the FERMENTER to a situation where it will sit at the correct temperature.

The other possibility is that you stored the FERMENTER with a lot of SODIUM METABISULPHITE inside, and forgot it was there. This will kill the YEAST, and ruin the taste of the brew. In this scenario, throw it out and start again. Obviously, exhaust the other possibilities first.

Scenario Three: *Brown Froth has erupted through the AIRLOCK.* This would most likely happen within the first 24 hours or so of adding the YEAST, The cause is almost always that you have over-filled the FERMENTER. Remove the lid, wash it, and replace. Should it happen again, repeat the process but use the tap and remove some of the WORT. No great problem will be caused to the finished product unless an infection got in.

   Scenario Four: The AIRLOCK is STILL POPPING after Seven Days or more. If the AIRLOCK is still popping at intervals of two seconds or more the most likely explanation is simply that it has been cold. Cold temperature slows down the fermentation process. It does not occur is the WORT has been kept in the correct temperature range. Before opening the FERMENTER, drain off some of the WORT into a jar, and check the Specific Gravity of the fluid with your HYDROMETER. If it reads 1.008 or more, leave it a few days. If it reads around 1.003 and you can see no continuous fine bubbles rising in the test jar that has been sitting for a few minutes, it is OK to bottle. Some SUGARS have additives that do not ferment out completely. (another reason to stick with common white sugar till you are more experienced) If you use an odd sort of sugar, your SG may never drop below 1.005. Sometimes, particularly in cold weather, there seems to always be a bubble rising in the AIRLOCK, even when fermentation is complete.

   Scenario Five: At Bottling Time things are not right. You open the FERMENTER to find the surface of the WORT is covered in an oil-like slick or film, with large bubbles rising. You open the FERMENTER to find a thick, even, white mold-like substance. You open the FERMENTER and there is a vinegar-like smell.

   The cause of all the above tends to be wild-born YEASTS and/or BACTERIA that has gotten into the WORT. Usually, this is because of a leak, or the lid being opened and having a random piece of bad luck in the air. Another cause is that the water in the AIRLOCK was low, and air got in at that point. Obviously, a lack of cleanliness and forgetting to sterilize will be culprits, and the only other possibility is that you used an infected YEAST.

   Dealing with the thick, even, whitish mold-like substance: To cheer you up, some of the very best of my brews have had this. (I may have forgotten the bad ones) Do not automatically throw it out, it may still be good. Take a slotted spoon or an egg-lifter, and scoop out as much of the surface film as you can. Then, with clean hands, take sheets of paper towels and lay them over the surface of the WORT. Lay them, one after the other, over the surface. The remaining mold will largely stick to the towels. When most of it is removed, go ahead with the bottling process. Discard the paper, of course, and make notes about it in your brewers handbook.

   The oil-like slick or film with the big bubbles and greyish color is difficult. It may be better to toss that attempt and start a fresh batch. I have tried to salvage the odd one like this, and the odd reasonable brew has resulted in my efforts, but only after I have gotten all the film out of the WORT. It is difficult to remove, but if you do make the attempt, only bottle 24 bottles, and leave plenty of room for the contaminated area of the WORT to stay out of your bottles. I have never gotten a really GOOD brew with this scenario.

   The Vinegar Smell generally indicates considerable exposure to air, and if this is the case it really is a vinegar and undrinkable. A very mild smell may be salvageable, and remember ALL WORTS have a slightly acid smell and are sourish to taste. You have to decide if the level of vinegar taste has gone too far. Once more, bottle only 24 bottles in this case.

After reading through all the troubles that can arise before bottling (I stress, these rarely happen with proper maintenance and management) please remember that the AIR is full of tiny microbes and wild yeasts. These microorganisms thrive in the sugary feed source that is your WORT, and they will do everything in their power to get in there. The solution is simple, be absolutely careful with cleanliness and sterilization, and do not remove the lid of your FERMENTER prior to bottling, even for a few seconds, unless it is for topping up.

Once more, this is as good an argument as we can raise for larger FERMENTERS, which would entirely eliminate the need for TOPPING UP, thus removing the need to remove the lid.

I have asked on many occasions if other home brewers have had the aforementioned problems. Of all the ones who have said 'no' in every case they were brewers who never took the lid off their Fermenter. Ignorance can be bliss! However, that said, after tasting their brews I honestly do not believe that they are as good. It is entirely a matter of personal choice, to TOP UP, or to leave it run. I like to TOP UP, because I can see the condition of the WORT at the mid-way point and, if there is a concern, I can correct it.

PROBLEMS AFTER BOTTLING

Scenario Six: The whole batch is dead flat after 14 days. You open the FERMENTER and the WORT is completely flat, with no bubbles at all. The most likely culprit is you, forgetting to add the PRIMING SUGAR. If the PRIMING SUGAR has been added, the next most likely culprit is TEMPERATURE, specifically if it has been extremely cold. In this case, putting your bottles in a warm place will kick the secondary fermentation into gear. At the correct temperature, gas will form up in the bottles during the first four days, and dissolve into the beer over the next week or so.

If the PRIMING SUGAR has been forgotten for the ENTIRE Batch, no problem, open them up and add it, then reseal. The same crown seals can be used if you remove them gently. If it is one or two bottles, you have to ask yourself if you want to do the lot again while you test for good ones. Your call.

Other possible causes are that the brew has been left too long and that too little YEAST remains in suspension, thus the PRIMING SUGAR has no action. You would have had to leave the WORT in the FERMENTER for over two months, but if so, open up the bottles, add a granule of yeast, and see if it kicks in after resealing.

Too many FININGS (clearing agent) may have been added, similarly removing all the YEAST. The WORT could have gotten TOO HOT and killed off the YEAST. Too much SODIUM METABISUPHITE may have been left in the BOTTLES, thus killing off the yeast necessary for the secondary fermentation. In this case, the brew may have been ruined, and you will never know the causes, unless, of course, you have kept a detailed log of everything you have done in your BREWERS HANDBOOK. If you look back, you may find your memory is triggered as to the likely cause.

As an example: Tuesday 7th October 2013. Sterilized all bottles with bleach and packed them away ready for bottling. You 'may' recall that you forgot to wash them out properly, because a phone rang, etc.

Scenario Eight: *The Whole Batch is Flat-ish, but has a Little Life* Most likely cause is simply that not enough PRIMING SUGAR has been added. Perhaps your teaspoon is too small, or you used a FLAT level teaspoon, which is not enough SUGAR. Too little SUGAR produces the above symptom. Your medium-sized teaspoon needs a ROUNDED teaspoon of SUGAR added. Even HEAPED works, but too little ruins the beer. Now, if you don't want to throw away your flat beer, mix it with a 'gassy' one. Remember too much SUGAR can cause your bottles to become too "gassy" and too little makes them come out flat.

Scenario Nine: *Odd Bottles in the Batch are Flat* You may have missed the PRIMING SUGAR with the odd bottle, or the seal may not have been good. If you find the common problem of a chipped rim on the bottle, this is the cause. You may have hit the crown seal too hard, or it may have already had a fault, it doesn't matter, throw the bottle away. You don't have to throw away the beer. As mentioned above, you can mix it with a good one. You "can" put more priming sugar into it, and there is every chance it will pick up, however, so in this case decant into a good bottle and repeat the bottling process.

Scenario Ten: *Beer remains Cloudy After Storage* Beer should be clear within two weeks of standing and bottling, and thoroughly bright and golden inside our weeks. (even without the use of a fining, or clearing, additive) If it is not clear, it was probably bottled before fermentation of the WORT in the FERMENTER was complete. Another potential culprit is that if you introduced grains into the WORT. This can create a STARCH HAZE (all grains contain starch). Poor quality YEAST can also be a culprit, but if it tastes OK there is no reason for not drinking it. There are also air-borne organisms that can cause this reaction, but you will taste it. NOTE: If the beer tastes 'off' for any reason, DO NOT DRINK IT!

Scenario Eleven: *Beer has an Unpleasant Taste* Sometimes you crack a bottle and the contents are just not right. The Beer has an unpleasant, bad, bitterish or lingering taste that is unpalatable. Check around to see if the Gods have cursed you. (In Ancient Egypt, Beer was considered to contain a spirit or a God. An 'off' bottle meant you may have been cursed.) In the modern day, it is unlikely, but you never know.

Beer should have a clean, fresh, pleasant taste. The reason for a problem like this can be manifold. The first, impure ingredients. Too many hops can cause a bitterness, but not a foul taste. If you have used old ingredients, they may have been contaminated. Alternatively, you did not properly boil the WORT before putting it into the FERMENTER. Proper BOILING ensures the ingredients are sterile.

You have made a mistake somewhere along the line because if everything is done correctly you simply CANNOT reach a foul tasting beer at the end. Did you sterilize properly? Did you remove the cleaning agents? Did you leave the lid off for too long when TOPPING UP? If there were no strange growths in the WORT before bottling, the most likely culprits are faulty ingredients or cleaning agent left in the WORT. DO NOT DRINK THIS BEER! Toss it out.

That said, you have not invested millions of dollars to make a sterile factory. Contaminants can get in if you forget to cover every step properly. If you stick closely to the Eleven Step Program I have outlined in this book, these sorts of problems will occur very, very rarely. Beat yourself up with guilt, and carry on.

Now, the beer may taste just a little bit 'off' which can happen if an airborne contaminant fond it's way into the WORT. This is generally safe to drink, and if you do a 'shandy' which is half beer, half lemonade or ginger ale, the beer generally is palatable. Even better, save it until relatives you don't like come visiting, and proudly pour it for them, saying you made it yourself!

Some people I have met habitually make bad tasting beer, but their taste buds have become so accustomed to it that they think it is great! Nothing we can say, other than good luck. They only have themselves to please, after all.

Variation: YEAST BITTEN: William Black wrote about this in 1835. There is a variation to this last section, where a beer can become what we call "Yeast-bitten". This means it has a clear and distinct smell of YEAST when you open it. The smell is unmistakable, something like freshly baked bread. It makes bread tastes great, and beer taste absolutely horrible, Discard it.

It only happened to me once, from too much YEAST in the FERMENTER. I had bottled my thirty bottles and wanted to quickly knock over the next batch, so I left all the dregs in the bottom of the FERMENTER and boiled up another WORT. I put it right in as soon as it cooled off. Well, the results were surprising, the fermentation cycle was fully complete in just 24 hours. THAT beer tasted good, and I felt at the time I was edging closer to the golden goal of "instant Beer".

So I did the same thing AGAIN, pitching a new WORT on top of the dregs, but I forgot, the yeast had grown and was now effectively many times the original. The first lot of "Fast Beer" was excellent, the second lot was absolutely foul, yeasty and sour. I had broken one of the primary rules, I had not sterilized the FERMENTER, and I had not followed through the correct steps.

NOTE TO SELF: There's no such thing as INSTANT BEER!

Scenario Twelve: *Before opening a bottle, you see a 'ring' inside the neck* This is not necessarily bad. A ring at the top of the fluid level inside your bottle does not make the brew a fail, but it does indicate you missed something. Usually, you didn't properly sterilize the bottles. If there is MORE than a 'ring' on the glass, like a film over the top of the beer, there has been an infection. It may still taste OK. The Golden Rule, if it tastes OK it probably IS OK.

But the message is to take more care in cleaning your bottles. Primarily you need to clean off this 'ring' before you reuse, and a good way to do this is to put 2cm (1 inch) of BLEACH in a washed bottle, and vigorously shake up and down, holding your thumb over the top. This dissolves the ring. Rinse until the smell of bleach is gone. *NOTE: Always begin with clean, shining bottles!*

Scenario Thirteen: *Beer Loses Its Head and Gas after pouring* The cause here is most likely the glass you are using. Household detergents not fully washed off, or the slightest amount of grease in a glass, can cause this problem. Detergent and grease are the enemies of Beer. Professional establishments clean their glasses every month in a bath of ACETIC ACID, which leaves your glasses with a truly sparkling clean appearance.

Another issue can be simply that you opened it too early. Your Beer should sit for a minimum of two weeks, and three months is more ideal. The issue here is that the Beer is not 'conditioned' (matured) and in such an instance it will lose its head and gas quickly. Remember, if the storage area has been extremely cold, this can also cause a very slow secondary fermentation.

Overall, I recommend you keep your Beer drinking glasses away from the kitchen and the general washing up liquids. Use a proper bottle washing compound and rinse them thoroughly. But little things, like eating greasy food then drinking from your glass, can put a greasy film into your glass, and flatten your Beer. As always, common sense is your friend.

Scenario Fourteen: *Beer Lacks Body, and Tastes Thin* This is usually a symptom of insufficient MALT in proportion to the other ingredients. Classic case, someone goes for a higher alcohol level by adding more sugar and less MALT. This ruins Beer. Too much SUGAR. Too much WATER. Not enough MALT. It will be one or more of these three.

Do not be afraid to add more MALT. You can make perfectly palatable Beer without any added SUGAR at all, just using the MALT and the HOPS. You can even use MALT as a priming agent instead of white sugar. I have mostly experienced a 'thinner' Beer when using the Light Dried Malt Powder instead of the LIQUID MALT EXTRACT. In this instance, increase the amount of the Light Dried Malt you are using to compensate for the lack of taste, and add a CARAMELISING AGENT. A dessert spoon or so of caramel color (Parisian Essence) suffices, but dissolve it in water first before adding it to the WORT.

Scenario Fifteen: *Bursting Bottles* There are FIVE principle causes for this:

- Bottling before Fermentation is finished
- Too much PRIMING SUGAR in the Bottle. (Maybe priming and forgetting, and doing a bottle twice - or too much in all of them)
- Overfilling Bottles. Not leaving a sufficient Air Gap at the Neck (The bottleneck, not yours)
- Weak Bottles (Beyond your control unless you used the thin wall 'stubby' or short bottles)
- Dropping the Bottle on Concrete. (A very reliable method of breaking them, but devastating to the nerves because of the stress you will feel from the deep loss you suffered)

Be Warned! One bursting bottle can set off a chain reaction, bursting the one stored beside it, and setting off a domino effect. I have heard the "Pop, "Pop", "Pop" of dozens of bottles each going off, one after the other. Bursting bottles can be lethal hand grenades, throwing shattered glass across the room at high velocity. USE CAUTION. If one goes, remove all bottles with that date on them, and wear protective clothing and eyewear as you do.

There is no solution to the bursting bottles, other than identifying them. Go to your BREWERS NOTEBOOK to find a likely cause, and seek not to repeat it.

Scenario Sixteen: *Bottle Erupts with Froth when Opened* Assuming that the bottle has not been shaken up by one means or another, there are two probable causes for this problem. Either too much PRIMING SUGAR has been used, or bottling was done before Fermentation in the FERMENTER was complete.

I have occasionally encountered this problem, and checking my Brewers Handbook have clearly identified that neither of these issues were the cause. In this case, I have presumed that some sort of infection got into the brew, and in this instance, the Beer had a bad taste, so it was almost certainly what the problem was.

Beer that explodes when opened like this can be virtually impossible to pour. However, if you wet the inside of your glass with water, it will settle it down. Letting the bottle stand with the top off will soon see an end to the eruption, but the Beer itself will tend to be a little flat by this time. Usually, it is still drinkable, however.

Scenario Seventeen: *Beer lacks Colour* This is almost solely due to a lack of MALT. It can also be because you are using LIGHT MALT, and need to add a caramelizing agent. A dessert spoon or so of caramel color (Parisian Essence) will suffice, but dissolve it in water first. The overall solution is often to use some of the DARK MALT EXTRACT. However, LIGHT MALT is excellent if you want the ultra pale looking beer. Taste will vary with the different types of MALT you use, and this is a matter of experimentation to find what you prefer.

If you have allowed bottles to be exposed to open sunlight, it can 'bleach' the contents. Again, check your BREWERS NOTEBOOK, and you find an answer.

BOTTLES

This deserves a chapter unto itself. Sourcing bottles, choosing the right type, the container your home brew comes in is important. There are also State Laws regarding recycling of some bottles, and you will have to check this out for yourself.

EDITORS NOTE: Google did not exist when this book was written. If you check online, GUMTREE and similar "Trading Post" type sites have no shortage of people selling empty 750 ml Beer Bottles from Eighty Cents to a Dollar each.

There are laws about certain types of bottles being re-filled, but to my knowledge, this does not extend to the conventional 730 ml Beer Bottle (AKA: Long Neck). Of note, the growing trend towards "Twist Top" bottles. This type is not convenient for home brewers. The simple bottles that use the lever-off bottle caps are the easiest and best to use. Before acquiring bottles, always check that they have the rounded lip, not the screw thread.

At the high point of my homebrewing career, I had over four thousand beer bottles. I lurked at garage sales, looked up flea markets, and sniffed about in my job as a real estate salesperson. I discovered that many houses have a carton or three of empty 750 ml beer bottles under them.

(Editors note: I can attest to Geoff scavenging ability. I find it so funny now to remember the look on his face when he found a cache of empty bottles. It was like he found gold! He would hold them up, inspecting each and every one for suitability, looking to see how much cleaning was required, etc.)

Please note that the 'no return' type of bottle often has a thinner glass wall than the old-fashioned ones and care is necessary when both handling and bottling this type. I recommend you simply do not use them, because one slight mistake of too much SUGAR, and off they will go, exploding everywhere This not only wastes beer, it creates a significant health hazard. You have to be extremely careful capping them as well, as they can crack when hammering on the crown seal.

Dark glass bottles are preferred at all times because too much light is the enemy of home brew.

That said, even clear glass, screw-top bottles CAN be used. They don't look the part, and they must be stored in near darkness, but it is not the bottle that maketh the brew. It just holds it all in place while the magic happens. In truth, the plastic drink bottles with the screw caps are so terribly convenient that nowadays it is hard to go past them as a brewing tool. With the screw top, they are also able to be stored easily in the fridge if you don't get through the whole bottle. They just have to be stored in a dark place, both for secondary fermentation and general storage, for best results.

All Australian cities have bottle recyclers. I presume this is true for most of the world. Sometimes you can do a deal with a manager there to pull out suitable bottles. Otherwise, you just have to hunt around and find them. I can assure you, there is no shortage of empty beer bottles at any given point in time and most drinkers will be grateful for you to take them away.

The BOTTLE is at the HEART and SOUL of your beer enjoyment, so it is more than just the UTILITY of the item, it is the APPEARANCE. Conventional 750 ml Brown Bottles look right and pour right. They are purpose designed for this specific job, which is why I recommend them.

I go over the cleaning of bottles in the chapter after the next but remember, at all times after you have found and cleaned your bottles, store them in a non-dusty, non-wind affected location, and store them UPSIDE DOWN.

Editors Note: There is another option available to home brewers instead of bottles, using the small, reusable "cask' that the micro breweries are now fond of. Look it up online if this is something you wish to consider. However, most would consider this applicable only for experienced brewers.

Example at: http://ournanobreweryproject.com/casks/

Standard 750 ml Brown Beer Bottle. Excellent reusable bottle for home brew purposes

"Stubby" type bottle. Thin walled, and not good to use due to potential for cracking.

Standard PET plastic screw top bottle. Suitable if stored in a dark place

"The beer tastes good to my throat, cold and bitter, and the three boys and the beer and the queer freeness of the situation makes me feel like laughing forever. So I laugh, and my lipstick leaves a red stain like a bloody crescent moon on top of the beer can. I am looking very healthy and flushed and bright-eyed, having both a good tan and a rather excellent fever."
Sylvia Plath

HOUSEHOLD BLEACH - The Brewers Friend

I t took me a long time to realize how much a benefit ordinary household bleach (Sodium Hypochlorite) can be in a brewer's life. This is especially true when you have to start out with dirty bottles you found under a house. You can find them with aged mold on the outside and inside, a thing that will defy the most strenuous use of a bottle brush.

To clean bottles in this condition, fill them with UNDILUTED bleach and let it stand for a few minutes. It will lift off the mold COMPLETELY with just some vigorous shaking of the bottle with the bleach in it, obviously covering the top with your hand. Now take your NEXT bottle, put in a funnel, and, using a teacup strainer to catch the solids, pour the bleach through this into the next one. Repeat the process with all your bottles.

When you are done, wash all the bottles thoroughly, until all scent of bleach is removed. Use your bottle bush as you do so to remove any loose sediment or contaminants still remaining. Bleach is such a powerful oxidizing agent that this alone is enough to sterilize your bottles. Note: You may still have a slight scent of chlorine remaining, but this may be from the tap water you have used.

Again, DO NOT STINT on STERILIZATION and CLEANLINESS. Almost ALL problems with home brewing stem from people being careless in this regard. You are using a LIVE yeast to FERMENT SUGARS and create BEER. Any bacteria, bug or wild yeast can affect the process, so remove as much as you can the possibility of these up front, and the quality of your brew will be superb.

That said, if you are NOT in the habit of cleaning your bottles with a bottle brush before and after every use, you will probably find a 'ring deposit' at the level to which the beer has been filled. A quick rinse with bleach removes this instantly and sterilizes the bottle at the same time, leaving it SHINY and CLEAN. Rinse it well, and STORE IT UPSIDE DOWN.

Let me stress the obviousness of this, do not store empty bottles right side up, because they will get contaminants and dust falling into them. Once sterilized with bleach and stored with the neck pointing DOWN the chance of bacteria or wild yeast entering is extremely slight. The bugs that ruin your beer are not particularly smart and don't go round looking for upside down empty beer bottles to climb up into. Gravity by its very nature will help keep your bottles clean. Again, the obvious, do not store them in dusty or windy conditions.

I use bleach to sterilize everything, including the FERMENTER, the LID, and the RUBBER SEAL, as well as the AIRLOCK. I also let some run through the TAP. It is SAFER than the SODIUM METABISULPHITE if there are asthmatics around as well. After the bleach, I simply HOSE DOWN all the components and have found this ENTIRELY SATISFACTORY for all my brewing needs. To revise: I scrub it all down, then put the FERMENTER back together, seal it, and add water to the AIRLOCK, ensuring it is ready to go for my next brew.

With the exception of the THERMOMETER, HYDROMETER (and it's jar) and perhaps the SPOONS and things of that nature, I rarely use SODIUM

METABISULPHITE anymore. For myself, I find it is perfectly useful, but it WILL NOT CLEAN DEPOSITS out of the INSIDE of a bottle, or the FERMENTER. Bleach is what breaks this up better than anything.

It is simply easier and cheaper to use household bleach for most of your sterilizing and cleaning needs. I have never had an "off" brew using it, despite the fact that my present brewing conditions (lack of space) are far from ideal. I have to brew in a garden shed and obtain my water through a garden hose.

Common, ordinary HOUSEHOLD BLEACH is your best friend. You can buy two liters for a dollar at any supermarket and please note that the expensive brands are OFTEN EXACTLY THE SAME PRODUCT as the cheap ones. But it pays to read the labels and buy the brand that offers the most of the active ingredient. You will see this written as "4%W/V Available Chlorine present as Sodium Hypochlorite" or "35g/ liter Available Chlorine present as Sodium Hypochlorite" or "Active Ingredient Sodium Hypochlorite - Available Chlorine 4%W/V"

In simple terms, the higher the number, the better. I often find the cheap versions are the same as the brand labels that cost twice the price. You DO NOT NEED the variations, such as 'lemon charged' etc. Chemists sell a brand called MILTON, which is used for sterilizing babies bottles. The label shows it holds 1% Sodium Hypochlorite, one-QUARTER of the cheap bleach, yet it costs five times the price. THE LABEL IS IMPORTANT, not the brand.

Bleach can irritate sensitive skin, so wearing of rubber gloves is recommended, especially when using it 'neat'. Old clothes should be worn because bleach will do what it does if you spill any on your clothes, and give you lovely large patches of white wherever it lands. Bleach SHOULD NOT be used with any other chemicals. It is an OXIDIZING AGENT and can react, sometimes explosively, with other compounds. It can also evolve toxic gases.

Obviously, wear eye protection. It hurts if you get any in your eyes. You are using it UNDILUTED in the bottles and FERMENTER, so you need to take basic precautions. If you have had a few bad batches in your home brew, start again from scratch and use BLEACH on everything. Sterilization, specifically the lack of it, is the most common mistake home brewers make. Just because it LOOKS clean is no measure of if it is sterilized.

STERILIZATION MUST BE CARRIED OUT WITH EVERY BREW! Before, on the equipment and bottles, and after when cleaning up. There are NO exceptions to this rule. When you, as the brewer, will grasp and accept this advice almost ALL of your problems will magically disappear. You will consistently produce excellent, fresh-tasting beer.

Household Bleach can justifiably be called the home brewer's best friend.

A woman drove me to drink ... and I didn't even have the decency to thank her.
W.C. Fields

BOTTLE WASHING

B e sure to read the PREVIOUS chapter on Household Bleach. If this method suits you, no more need be said. However, for the sake of completeness, I will give you the other main method of Sterilization, other than Bleach. Primarily, DO NOT USE DETERGENT in washing bottles. Even after repeated rinsing, it can and will still leave a film.

If you are using bottles from a previous brew, it is imperative to wash them as soon as possible after emptying, thus preventing contamination rather than trying to solve it later. Hot water is best, but ordinary room temperature cold water is fine. Rinse the bottle four or five times. Next put 13cm (5 inches) of HOT water into the bottle, and using the bottle brush clean it well, then rinse.

Then sterilize each bottle with SODIUM METABISUPHITE. As mentioned, two heaped tablespoons per two liters of water is the correct ratio. I recommend that you have a prepared solution on hand, as it is entirely reusable for many purposes. Put 13cm (5 inches) of the solution into the bottle, cover the top, and shake vigorously. Decant the fluid back to the holding container, and rinse the bottle. NOTE: If you leave SODIUM METABISULPHITE solution in the bottle without rinsing it can create an unpleasant taste in your beer.

Most direction for using SODIUM METABISULPHITE involve soaking things in it, and they recommend half the strength I am suggesting. It is very time and space consuming to leave things to soak, and most could not be bothered. Use a stronger solution, but it definitely needs rinsing.

Some brewers do not believe in rinsing after sterilizing. They have the opinion that the tap water can contaminate the bottles. Well, as most of these same brewers USE tap water when preparing their WORT and TOPPING UP their beer, this is a spurious argument. Rinsing and storing UPSIDE DOWN is recommended. Baseline, if you drink the water from your tap, it is fine. If you don't, and you are using spring water because it is too contaminated with fluoride or chloride, then rinse with the spring water or boiled water. Treated city and town tap water is unlikely to create problems, but if you believe it to be suspect, boil it first.

SODIUM METABISULPHITE is widely used as a preservative in processed foods, Unless you are asthmatic, there is rarely an instance of complaint using this product. If you ARE asthmatic, you must use bleach for sterilizing, and you must rinse. (As ordinary bleach can also create an issue)

Your SODIUM METABISULPHITE solution can be used many times before it is rendered inert. When the strong smell is gone t is time to make another batch. DO NOT take a deep sniff to check, it is like ammonia. When transferring from one bottle to another, it is a good idea to use a tea strainer, to remove any solids.

If you are starting with fresh, clean new bottles, great. But most of the time, the old ones have a crust f something inside them, and the use of bleach, as noted two chapters previous to this, is the best 'solution' - pardon the pun. Again, when using bleach to clear out old bottles, use a strainer to remove mold and solids.

YEAST

Yeast is a member of the plant kingdom. As it has no trunk, leaves of branches, it is classed as a FUNGUS. There are over 80,000 known varieties. Technically speaking, it is a single-celled fungus (Saccharomyces) which grows and multiplies via single-cell division. Being a living organism it needs to feed and have a suitable environment.

We are only concerned with beer yeasts, of which there are two varieties - Top-Fermenting and Bottom-Fermenting. Top-Fermenting (Saccharomyces Cerevisiae) thrives best where there is air available, as in an open Fermenter. Bottom-Fermenting (Saccharomyces Carlsbergensis) work best at the bottom of the Fermenter, in the absence of air.

It is no coincidence that the name looks like Carlsberg Beer because these are the people who discovered this yeast. **Editors note** Carlsberg Beer: Founded in 1847 by J.C. Jacobsen and named after his son, Carl, **Carlsberg** created a pure strain of yeast in 1883 that completely revolutionized beer brewing.

To grow vigorously and do its job of fermentation properly, YEAST needs SUGAR of some kind, or a starch converted to a SUGAR, plus, surprisingly, minerals and vitamins. Malt is an ideal and complete food for YEAST. Honey is as well, but refined sugar is not. This is a pure carbohydrate without vitamins or minerals. YEAST will not ferment a WHITE SUGAR solution on its own but will do so if sufficient MALT is added. YEAST CONSUMES SUGAR and converts it to ALCOHOL and CARBON DIOXIDE GAS. While it is doing this, it is growing rapidly itself.

In other words, you feed your YEAST the right food and it gives you booze. Very good of it, really.

When you place your WORT into the FERMENTER for the first time, you add your tiny packet of YEAST granules. After this, it grows, and by the end of Fermentation there will be half a cup of more of pure YEAST residue inyour FERMENTER. Just a teaspoon of this is sufficient to start the fermentation process in your next brew. If the beer it has produced looks good, there is no reason not to re-use the YEAST. Obviously, if there have been an issue, consider just tossing it.

In the beginning, you will not be certain if your brew has worked out well. Take the YEAST produced and put it into a sealed container in the fridge or freezer until you have had time to evaluate the beer it came from. If the beer is good, keep reusing that yeast. There is absolutely no need to spend money on more brewers yeast unless the batch it produced was off.

YEAST is high on the survival scale. It can be dried into granules, which gives it convenience and resistance to heat whilst in storage or transit. It can be frozen for long periods and still be fully active when thawed. What will kill it are chemicals and too high a temperature.

Just how high, I don't know, but in an experiment with one brew, I raised the WORT temperature to 45.5 C (114F) by using hot water, then pitched (added) the YEAST in a closed Fermenter. To maintain a warm temperature, I covered the Fermenter in a box with a 100 watt light bulb burning inside.

The next day, fermentation was well underway and the WORT temperature was 38C (100F). Fermentation was COMPLETE in just 48 hours and the beer turned out to be excellent. The YEAST I used was from a previous brew, which had used the dregs from the bottom of a bottle of Coopers Sparkling Ale. This is worth remembering if you live in a very hot climate and are put off by claims that YEAST is killed in temperatures over 35.5 C (96F). There are not many localities where temperatures exceed 45.5 C (114F) bearing in mind that the WORT temperature is generally lower than the ambient air temperature. Yeast granules should perform similarly, especially a top-fermenting type.

The Experts say the ideal heat range for Bottom-Fermenting YEASTS is around 17C to 18C. (63 - 65F) Above this, they say, many yeast cells are destroyed and you can get an 'off' taste. They should know their business, but I can say I have been more than happy with beers produced at higher temperatures. *I suspect this temperature variation is more critical with wine production than beer.* Mind you, I have kept a particular strain of YEAST active and reused for many years. Its quality has remained excellent even with a wide range of temperature fluctuations.

Most of the time, in a closed Fermenter, we are using bottom-fermenting yeasts. However, you CAN use open Fermenters which call for a Top-Fermenting YEAST. These work on TOP of the WORT, as would seem obvious, and thrive in the presence of AIR. They produce a thick brown or a white frothy mass on top of the WORT. Most authorities advise you to remove this from time to time. But Coopers, who consistently produce excellent BEER PACKS, advise otherwise.

They advise to use OPEN CONTAINERS and advise strongly against removing the evolved mass at the top of the brew. I tried one of their bitter packs using the open method without skimming and the beer was first class. Surprisingly, residue on the bottom of the Fermenter was no different to that of a closed Fermenter.

There is one significant difference in the process: ***Beer in OPEN FERMENTERS should be bottled as soon as possible after fermentation has ceased.*** During fermentation, the covering of CO_2 gas (a heavy gas) helps protect it from infection, but when fermentation ceases this evaporates. Infection by wild yeasts or bacteria can occur. This does not happen with a Closed Fermenter.

We have discussed Top and Bottom Fermenting YEASTS. Some suppliers have a bit of an each-way bet and market a YEAST they describe as suitable to both Top and Bottom fermentation processes. To me, it sounded more like wet and dry sandpaper, or sweet and sour sauce!

On one occasion I was experimenting with a bottom fermenting YEAST that came from a commercial brewer. I did all the usual preparations for the WORT and added this commercial brand yeast. On the following day, the AIRLOCK was popping every three seconds, indicating a very active fermentation. Against the rules, I opened the top to take a look.

The scene was fascinating to watch, as there was NO Froth or Scum. The surface of the WORT was clear. But there was a rolling, boiling disturbance from the bottom of the WORT, and the whole thing was seething. It was something entirely unique and different. I have a deep gratitude for YEAST, it give me beer!

ADDING FININGS

Editors Note: *The original chapter was called 'Finitration" but it has gone missing from the manuscript. I have put this section together from collecting various notes.*

You ADD FININGS the day PRIOR TO BOTTLING in order to reduce cloudiness in your final product, caused by an excess of suspended particulate in your beer. The most common FINING is simple GELATIN POWDER, a teaspoon of which you mix into warm water, then just pour into the WORT after fermentation is complete. You do this the day before you bottle.

There are a number of issues, the primary one being you are opening the FERMENTER at a critical time, where there is no CO_2 evolvement on the surface of the WORT to protect it from bugs. Most people recommend you stir in the FININGS, but did they sterilize the spoon? It is an area which can and does cause problems with your final product.

In simple terms, if you have done everything right, sterilized everything, kept accurate records, observed good and bad yeasts and taken steps to remove faulty versions, then cloudiness in your beer is not a problem. I almost NEVER added any FININGS, and rarely, if ever, had a cloudy beer. Rather than go into great lengths explaining how to use the various types that are on the market, I would prefer to push you more towards accurate record keeping and attention to detail.

In my humble opinion, it is more important to understand WHY you are getting the cloudiness than covering it up with FININGS. This is covered in the TROUBLE SHOOTING section some chapter prior, but USUALLY, the culprit is people introducing a GRAIN for FLAVOUR, and creating a STARCH HAZE in the final product.

We are not covering beers made from Grain Mash in this book, and sticking to the basics, but there are many people in the brewing shops who can help and encourage you if this is the direction you wish to go in. It is considered to be the "real beer" when you start and create the ingredients yourself from scratch and, once you get underway and develop a little confidence, I strongly recommend you give it a go.

I got my cost per bottle down to under TWO CENTS by creating everything myself, so you might say, I wrote this book to give you my two cents worth!

Editors Note: *The internet happened since this book was originally written. There is no end to the advice you will get on on-line forums, and there are some incredibly knowledgeable people out there. However, for the most part, there is mostly people with limited experience and multiple opinions, and most of these are wrong. If you stick to the basics as outlined in this book, you will never be led astray. Everything Geoff suggests comes from trial and error, and many years of experience.*

The BEER HYDROMETER

The Beer Hydrometer is an instrument made of glass or plastic which, when floated in liquid, measures its SPECIFIC GRAVITY (SG) or DENSITY. When SUGARS are added to water, the water becomes more dense and the Hydrometer floats higher. The more sugar, or sugars, that you add, the higher it floats. In beer brewing, the Hydrometer has two purposes.

- It will indicate when Fermentation is complete and the beer is ready to bottle (around 1.003 of lower, and as close to 1.000 as possible)
- It will enable you, by using a simple formula, to calculate the PERCENTAGE OF ALCOHOL in your beer. You take one reading BEFORE adding YEAST, and another reading PRIOR to BOTTLING. (If TOPPING UP with added water during the fermentation cycle, an allowance has to be made for this)

The beer Hydrometer is calculated to read 1.000 in pure water, with a temperature of 15.5C (60F). If the temperature varies by more than 2.8C (5F) either side of this figure an allowance must be made. (See table below)

The following table was calculated by testing Gold Coast city town water at the temperature indicated. It serves to give you an 'allowance factor' when temperature of the WORT is taken into account against the sugar/alcohol level in your brew. Note, the temperature of the WORT is what matters, not the surrounding air.

TABLE A

Degrees Celsius	Degrees Fahrenheit	Hydrometer Reading	ADJUSTMENT
4.5	40	1.002	-0.002
10	50	1.001	-0.001
15.5	60	1.000	correct
21	70	0.999	+0.001
27	80	0.998	+0.002
32	90	0.996	+0.004
38	100	0.995	+0.005
43	110	0.994	+0.006
The right-hand column is the TEMPERATURE VARIANCE FACTOR you must adjust for			

Please remember that the above are temperatures of the liquid in which you are floating your hydrometer, and have no relation to air temperatures. As the temperature of the liquid increases, the level of the Hydrometer lowers, so it is necessary to ADD the applicable figure to your Hydrometer reading. Remember, a negative figure added to any number LOWERS that number.

So, take a measurement with your Thermometer of the WORT, and match it to the right-hand column of this table. When you get your Hydrometer reading, you ADD this figure to whatever it says, and you will have a properly adjusted reading.

Example One: Your Hydrometer reads 1.035 and the WORT temperature is 35C (95F). Add 0.004 and you get the CORRECTED reading of 1.039

Example Two: Your Hydrometer reads 1.005 and the WORT temperature is 7C (45F). Add MINUS 0.002 and you get the CORRECTED reading of 1.003

Example Three: Your Hydrometer reads 1.003 and the WORT temperature is 18C (65F). There is NO VARIANCE so the CORRECTED reading is 1.003

There are various makes of Hydrometers, and a wine Hydrometer is perfectly suitable. They are all largely marked out in the same way. Mine is a RITCHIE brand and is marked at 0.990, then 1.000 (Zero mark when floated in pure water at a temperature of 15.5C - 60F) It then goes to 1.010, 1.020, 1.030, etc.

The conventional way of recording readings above 1.000 are to use only the LAST TWO DIGITS of the reading. 1.003 become 03, as an example. 1.004 becomes 04, 1.040 becomes 40, and so on. In practice, your HIGHEST reading will occur prior to pitching the yeast, before all the SUGARS begin to convert to alcohol. The LOWEST reading will occur when FERMENTATION is complete.

If your FIRST reading is 1.039, this is called the ORIGINAL GRAVITY (OG) of 39. If, when fermentation is complete, the reading is 1.003, this is called the FINAL GRAVITY (FG) of 03 (Assuming in both cases you have made adjustments for temperature). Your BREWERS HANDBOOK should have the notes. Brew 05/07/1978 (Recipe used) OG 39 / FG 03. *This is tremendously important information to retain when you are tasting your final outcome.*

Be particularly careful regarding the temperature of the WORT at bottling time. Some instructions tell you to bottle when the Hydrometer reads 1.006 or below. However, if your WORT temperature is 32C (90F) when 04 is added to your reading it becomes 1.010 which is likely to give you a very poor beer, along with bursting bottles if bottled at this stage.

A point that needs to be made here is that some SUGARS will never ferment out fully. Lactose (milk sugar) which is sometimes used in the WORT to sweeten STOUT cannot be fermented AT ALL by beer yeast. It will raise your Original Gravity, and when fermentation is finished, it will also raise your Final Gravity reading by that same amount. Glucose, or brewing syrup, is another product which contains some unfermentable sugars. Obviously, if a lot of GLUCOSE has been used in the WORT, it will throw out all your Hydrometer readings.

Hydrometer readings are best done using a straight-sided glass jar and done at eye level. A milk bottle may be used if your Hydrometer is not too long. If your Fermenter is full almost to the brim (after topping) a reasonable reading can be taken there and then in the Fermenter itself. When the level is low, such as before pitching the Yeast, it can prove difficult to get an accurate gauge. The reason for being at EYE LEVEL is to ensure you are reading the right numbers.

It will be noted that when the Hydrometer is floated in any liquid, the fluid tends to rise up the stem. Allowance must be made for this and the reading taken at where you estimate the true level to be. A straight-sided glass test jar with a base for it to stand up in is a good investment.

There is a strange variation to Murphy's Law when it comes to Hydrometers. They are only calibrated on one side and for some peculiar reason, this side always seems to spin around and face away from you, making it difficult to read. If you have a test jar standing on its own pedestal on a circular table this is no longer a problem. Murphy CAN be defeated with a mirror or a sprint around to the other side of the table.

Sadly, I don't have a circular table, but even if I did, I am quite certain the wretched thing would go out of its way to turn around as soon as I got to the other side. Now, you could be smart and rotate the test jar but the intransigent, intractable Hydrometer will remain facing the way it was. Also, if there are any bubbles at all in the air, they will leap over to obscure the view of any reading you try to take.

You can test your Hydrometer by floating it in water at the temperature of 15.5C (60F). It should read 1.000. I tried mine is pure Brady, an alcohol being less dense than water, and it sank to the bottom. In Sweet Sherry, it bobbed at a 1.036 reading. This gives a clear indication of just how much sugar is in sweet sherry. Semi-Sweet Sherry gave a reading of 1.009.

SUMMARY

When using a Hydrometer, temperature refers to the temperature of the WORT, not ambient air temperature. The Hydrometer reading should be checked against the scale given regarding temperature, and adjusted accordingly.

Your Hydrometer reading needs to be taken at the point you estimate the true reading should be, not with the fluid rising up the stem.

Bubbles affect the Hydrometer reading, tending to force it up thus giving a false reading. When decanting beer from the tap of the Fermenter, bubbles always ensue, so let the test jar stand for a few minutes before attempting a reading.

If you want to test previously bottled beer at some point to see if it has fermented correctly, or to check if you have some unfermentable sugars, note that it takes a minimum of two days for that beer to go completely flat. This is so even in an open glass. Any bubble activity at all can give a false reading.

Until you gain experience, always test your beer with the Hydrometer to ensure that fermentation has finished - being a Hydrometer reading of 1.006 or below, temperature adjusted. With a little practice, your eye will tell you and the only reason for the Hydrometer reading will be to calculate the percentage of alcohol.

Which leads nicely to the next chapter.

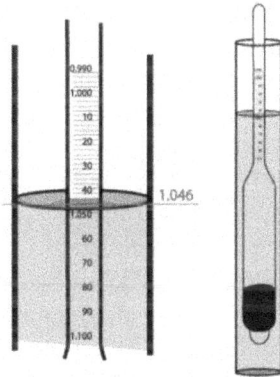

How to Calculate Alcohol Content

P lease be sure to read the previous chapter regarding the use of Hydrometers, and get familiar with using this instrument. It is one of the ESSENTIAL tools in brewing.

Let's say you took the Original Gravity (OG) of the WORT before you pitched in the YEAST. The reading on the Hydrometer says 1.039 (39) after allowing for temperature correction, so write this down in your BREWERS NOTEBOOK, together with the quantity of WORT in the FERMENTER.

When FERMENTATION is complete and you are ready to bottle, take another reading and make corrections is necessary for the beer temperature. Say the corrected reading in 1.003 (03). This is the FINAL GRAVITY (FG)

The formula to calculate the percentage of alcohol by volume is:
Subtract FINAL GRAVITY from ORIGINAL GRAVITY.

Next, divide this result by 7.4. In the case of the above, from the OG - 1.039 (39) subtract the FG - 1.003 (03) and you get: 39 - 03 = 36

Divide 36 by 7.4 and you get 4.86.

You have a beer with 4.86 percent Alc/Vol.

Now, remember, you will be adding SUGAR to the final bottling. If it is a rounded teaspoon, this will add approximately 0.5 % alcohol content to your brew. So ADD THIS to the 4.86 and you get a FINAL ALCOHOL reading of 5.36%. Obviously, if you are adding water to the BOTTLES and did not TOP UP in the middle of the process, this affects the Alcohol level. We cover this next chapter.

Are we starting to see how important note taking at every step is? Specific Gravity readings are just ONE of the important requirements of knowing what you actually have in your bottle. For instance, you have friends over and you pour them some home brew. You have three glasses each, and they say how great it tastes. Then they drive home, to be pulled over and charged for drink driving. Your home brew had a 6% alcohol content, where you thought it only had 4.2%.

The above calculation is simple and important. It gives you an accurate estimation of the percentage of alcohol in your brew. The percentage of alcohol is expressed as (example) "5% Alc/Vol". What this MEANS is that alcohol forms up 5% of the volume of the liquid in your beer. A 750 ml bottle at 5%Alc/Vol has 37.5 ml (1.3 fl oz) of PURE ALCOHOL.

People in the street will either call me 'Prime Minister' or 'Justin.' We'll see how that goes. But when I'm working, when I'm with my staff in public, I'm 'Prime Minister.' I say that if we're drinking beer out of a bottle, and you can see my tattoos, you should be comfortable calling me 'Justin.'
Justin Trudeau

How to Calculate Alcohol Content - (when water is added to bottles)

To be able to accurately project the alcohol content AFTER you add water at bottling for your Brew you must know the potential alcohol in your fermented product plus the exact quantity of fluid in your original WORT BEFORE you added water at TOP UP. In this case, you haven't TOPPED UP, so the amount of fluid in your original WORT before you pitched in the YEAST.

Your ORIGINAL GRAVITY is the Hydrometer reading BEFORE you Pitch in the YEAST. After making corrections for temperature, you write this into your WORKBOOK as well as the quantity of fluid in the FERMENTER. When using a standard 22.5 liter (5.9 US gallons) Fermenter, you would probably have 18 liters (four gallons) of WORT.

EXAMPLE: You have 18 Litres of WORT. The temperature is 32C (90F). Your Hydrometer reading is 1.036. Your TEMPERATURE CORRECTION requires you to add 0.004, which makes this figure 1.040. This is the figure you write down as your OG. (Original Gravity) It is customary in brewing to write down only the last two digits, so you write it as 40.

You are NOT TOPPING UP, so you wait for your next reading until BOTTLING TIME. Take another Hydrometer reading, again allowing for temperature adjustment. This adjusted reading is 1.003 (03) so now you simply subtract this last reading, 03 from 40 to get 37. We now divide 37 by 7.4 to get 5. You current projected alcohol content is 5% Alc/Vol. But now you are going to ADD WATER and SUGAR to the bottles before bottling. We need to take these variations into account.

As a note: Even if you have Topped Up and added the WORT up to 22.5 liters (US six gallons) by the time you take this final Hydrometer reading it makes no great difference. The original Gravity and the quantity of fluid in your WORT at the start of the process is all that matters.

In this case, we have NOT TOPPED UP and are ADDING WATER to each bottle, which will REDUCE the alcohol content. Take the 5% you have resolved, then MULTIPLY IT by the quantity of fluid in the WORT at the start, then DIVIDE it by the total amount of fluid you will end up with. If you are adding 150 ml to 30 bottles, you are adding 4.5 liters (1.18 US gallons) to get 22.5 liters (5.94 US gallons).

To work in Metric to make the numbers simpler: 5 x 18 = 90 divided by 22.5 = 4

Your ADJUSTED ALCOHOL reading is therefore 4% and by this calculation your friends will no longer be arrested by the police for drinking three glasses over 2 hours. THIS is why it is important.

HOWEVER, you are ALSO adding SUGAR to the bottle, and you must ADD 0.5% of additional alcohol produced by the SECONDARY FERMENTATION. So your REAL final reading is 4.5% Alc/Vol which is on the border of whether your friends get arrested or not.

NOTE on WORLD PROOF Systems

Important Note: This advice on alcohol content is a GUIDE ONLY. It is an accurate guide, but drinking and driving with too high a blood alcohol level is a serious offense in the eyes of the law and can get you banned from driving. Please err on the side of caution when drinking your home brew. In later years, I have tended towards LOWER ALCOHOL beers and stouts and found no loss of taste or pleasure from drinking. In fact, I can drink more before falling over, so this must be a good thing.

Another method of expressing alcohol content is "proof spirit' and there are three different systems for this, namely British, United States, and Metric (Gay Lussac Method).

BRITISH - 100% pure alcohol is 175 Proof

U.S.A. - 100% pure alcohol is 200 Proof

Metric (G.L.) - 100% pure alcohol is 100 Proof

100% alcohol is given in round figures as some impurities may reduce this very slightly.

To convert alcohol by volume into British Proof, multiply the percentage of alcohol by volume by 1 3/4. As an example, taking a figure of 5% alcohol by volume: 5 x 1 3/4 = 5 x 7 over 4 = 35 over 4 = 8 & 3/4 proof

To convert alcohol into U.S.A Proof, simply double the alcohol figure, which in the case of 5% Alc/Vol equals 10.

Metric Proof (Guy Lussac or G.L.) in degrees is the same as percentage by volume, IE: 5% Alc/Vol = 5% proof. (much simpler)

In Australia, most commercial beers (except the popular lights) are around the 5% Alc/Vol mark. Alcoholic spirits (Whiskey, rum, gin, etc.) range between 37.2% and 44% Alc/Vol. Fortified spirits, such as port, sherry, etc. are around 18%. Unfortified (no added spirit) table wines such as Moselle, Shiraz, etc. are around 9% to 14%.

Editors Note: https://en.wikipedia.org/wiki/Alcohol_proof

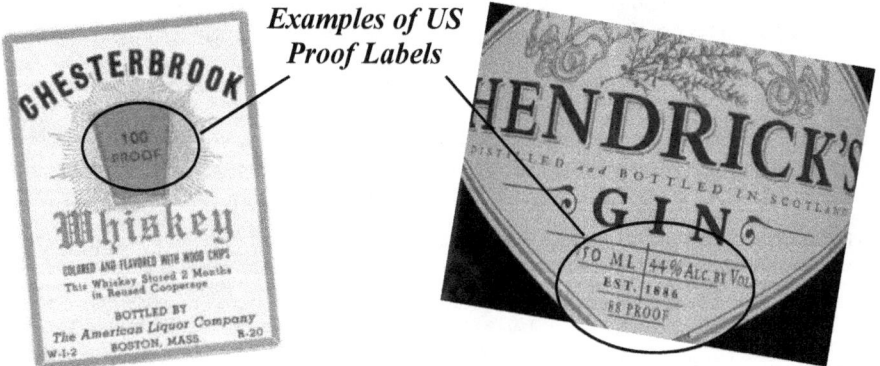

Examples of US Proof Labels

Calculating the AMOUNT of WATER to ADD to BOTTLES

You may wish to resolve the amount of water to add to a bottle in order to get to a specific level of alcohol in your final product. To do this you need to know how many bottles you want to finish up with after adding the water. You 'can' increase the number of bottles by a factor of six or more. Obviously, it all depends on the size and quantity of the bottles you will be using.

Example: You have NOT TOPPED UP your FERMENTER. You have 18 liters (4.75 US Gallons) of beer ready to bottle. You have thirty x 750 ml bottles (25.36 US Fl Oz) that you wish to use. Simply multiply 30 x 750ml to get 22.5 liters (5.94 US Gallons) and subtract your ORIGINAL WORT from this. 22.5 minus 18 = 4.5 ... 4.5 liters is the additional amount of water needed. Divide this into 30 bottles, and you get 150 ml per bottle.

Let us say you have 36 x 750 ml bottles. Again, same scenario as above, you multiply 36 x 750 to get 27 liters that you are aiming to fill. 27 liters minus the original 18 liters gives 9 liters. This means you need to add 9 liters between 36 bottles. I am using Metric because it is easier with the numbers. Here many people make a mistake and divide 9 into 36 to get 4. Rather than think it is 4 liters, which is absurd, they presume it must be 400 ml. This is INCORRECT.

The difference between the wort of 18 liters and 36 x 750 ml bottles 9 liters, or 9000 ml. This figure is then divided by 36. This equals 250 ml per bottle, the CORRECT amount of water that is needed per bottle.

Obviously, whatever number you come to that needs to be added to the bottles, you need a measuring cup to add that specific quantity to each bottle. Remember, you will also be adding PRIMING SUGAR, so you will want to ADD THIS FIRST because otherwise, the sugar will stick to the neck of the bottle.

You need a DRY FUNNEL for the sugar and a WET FUNNEL for the water. As previously mentioned, an offset rubber band around the WET FUNNEL will create an AIR GAP to allow air to escape the bottle when filling. Trust me when I say that these small details make a HUGE difference in reducing spillage and frustration at this last stage.

It took me years to realize how much EASIER pouring the final product into each bottle was by just adding the offset rubber band at the point on the funnel where it will sit in the neck of the beer bottle, so as to allow an air gap when pouring. I probably saved thousands of dollars in visits I didn't have to make to the chiropractor!

In every single step you make in home brewing, I promise you, small details make a huge difference. Things like that adding of FININGS (using gelatine to reduce cloudiness in your beer) can be sidestepped by the good recording of details, to find out if it is the YEAST, or AIR in the FERMENTER, or improper sterilization that is the cause of the cloudiness.

Ready Reckoner Table

This is a ready reckoner table, based on a final gravity reading of 1.003 and it includes the sugar added for the Secondary Fermentation process. It relates to what your specific gravity reading is in the WORT prior to the yeast being added, and can allow you to adjust your ingredients before any part of the process gets fully underway. (IE: If the percentage of alcohol is too low, you can add more malt or sugar)

TABLE B

Specific Gravity BEFORE Fermentation	Percentage Alcohol when Fermented
1.010	1.5%
1.025	2.1%
1.020	2.%
1.025	3.5%
1.030	4.1%
1.035	4.8%
1.040	5.5%
1.045	6.1%
1.050	6.8%
1.055	7.5%
1.060	8.2%
1.065	8.9%
1.070	9.5%

These figures are based on NO MORE WATER being added, either with TOPPING UP or to the BOTTLE at the end of Fermentation. Allowance for SUGAR ADDED in the SECONDARY FERMENTATION has been made.

The purpose of this table is simply to assist you to understand PRIOR to Fermentation as to where your brew is heading as far as alcohol content is concerned. To explain, the Original Gravity (OG) of your WORT reflects the density of the fluid, and it is the SUGARS that create this density. A higher OG tells you there is more sugar, therefore a higher alcohol reading will result.

Editors Note: There are extensive letters to Universities and notes to brewers regarding this subject. It also encompasses the 'type' of sugar used, as well as the temperature of the WORT. More details on page 74.

"Beer, if drunk in moderation, softens the temper, cheers the spirit and promotes health."
Thomas Jefferson

Are GLUCOSE and DEXTROSE the SAME THING?

Various books on brewing will tell you they are the same. If you look up a dictionary you are likely to arrive at the same conclusion, as they are both C6 H16 O6 and further they appear to be used interchangeably. They are also, confusedly, called Grape Sugar. You will also see DEXTROSE defined as Corn Sugar, Dextroglucose, Start Sugar and the sugar found in blood and many plants. I have asked several chemists what the difference as, and all of them did not know. The only way I could find anything conclusive was by contacts the manufacturers of the various products.

The fact is that Dextrose and Glucose made in Australia are NOT the same product for brewing purposes. They are both made from wheat starch, from which the bran, Pollard and wheat-germ have been removed. White flour is used and, by a process called Hydrolysis (treatment with water and acid and/or enzyme), the gluten or protein component is extracted, leaving pure starch. This is the same starch used in laundries, or which is sold as "Corn Flour" made from wheat.

This starch, by further treatment, is converted into glucose, a clear syrup, which can be produced in varying grades, with different characteristics for differing purposes.

The simple sugar dextrose (dextrose monohydrate - a monosaccharide) is a sweet, white powder. This is the final product from the whole starch-hydrolysis process. To my knowledge, though others manufacture glucose, there is only one manufacturer of dextrose in Australia, Goodman Fielder of Tamworth, NSW.

Glucose, however, is an INCOMPLETE hydrolysis of starch containing dextrose, maltose, tri and tetra sugars, and higher sugars with all having around 20% moisture. (It can be dried out to a powder.)

In brewing, while DEXTROSE is a fully fermentable sugar, GLUCOSE contains hydrolysis products of starch and higher sugars which cannot be metabolized by the yeast into alcohol.. As a consequence, these products can remain to affect the taste of your beer, and the starch, in particular, can contribute to cloudiness. In the positive, they can help 'head retention' due to the presence of protein and higher starch fragments.

In the U.S.A. (which produces most of the worlds corn) glucose (corn syrup) and dextrose are made from Corn by the same hydrolysis process. Incidentally, one State, Iowa, produces more corn than any other country in the world, excluding the U.S.A. *(Editors Note: Since this book was originally penned - 1987 - the relationship of Corn Syrup to obesity has made this a suspect product)*

In brewing, DEXTROSE ferments COMPLETELY, whereas GLUCOSE (brewing syrup) does not, thus leaving a sweetish taste in the beer if a lot of it is used. Dextrose can be bought in supermarkets under the brand name, Glucodin. Confused? It took a while to unravel the puzzle.

They are NOT the same thing, despite being derived from the same source. One ferments completely, the other ferments only partially.

POURING HOME BREW

This might seem to be an odd subject for a chapter heading, but a certain amount of sediment is unavoidable in home brewed beer. The quality of the YEAST (in a properly managed brew) will determine how sludgy the sediment is, or whether it sits as a firm matt on the bottom of the bottle. The latter allows the beer to be poured clear to the last drop.

Let us assume that, in fairly normal circumstances where the YEAST is not as good as it could be, that the sediment is of the sludgy type, which does not settle 'hard' and that the beer has been stored upright in the refrigerator.

The best method would be to gently decant the beer into a chilled jug, carefully leaving the sludge at the bottom. Of course, this way the beer goes flat far more quickly if there is only one person drinking it. The 'other' way is to pour the beer slowly into a tilted glass and gently return the bottle to the upright position. The less angle you put to the bottle, and the slower you move, the less disturbance to the sludge. Repeat for the second glass, and so on, stopping when you see the first sign of sediment.

If four or five glasses are to be poured, pour the first glass and, keeping the bottle tilted at all times, fill the others. Only return the bottle to the upright position when all glasses are filled. Being careful to not mix the sludge into the beer, you will find the brew will have clarity and an excellent taste.

However, let us not despise the dregs! Although you may not like the taste overmuch, if you swill the remaining beer in the bottle around with the dregs, and drank the lot, you are getting an excellent tonic that is a rich source of ALL the "B" vitamins, as well as amino acids and proteins. Plus, there is a rich abundance of readily absorbable minerals, including the essential Chromium and Zinc. Brewer's Yeast is one o the richest sources of Chromium, the so-called Glucose Tolerance Factor. It is also one of the main minerals depleted from your system by smoking tobacco.

If you have been fortunate enough to get a hold of a good quality YEAST (and you can make your own) this is far easier. It forms a MATT at the bottom of the bottle, and you can just pour normally. As a note: If the beer is for any reason too frothy to pour, just wet the glass or the jug and this will simplify pouring.

Which brings us to an often asked question, "Can I filter out the sediment from my home beer?"

"Whenever the devil harasses you, seek the company of men or drink more, or joke and talk nonsense, or do some other merry thing. Sometimes we must drink more, sport, recreate ourselves, and even sin a little to spite the devil, so that we leave him no place for troubling our consciences with trifles. We are conquered if we try too conscientiously not to sin at all. So when the devil says to you: do not drink, answer him: I will drink, and right freely, just because you tell me not to."
— *Martin Luther*

FILTRATION

The majority of Commercial Beers are filtered to remove all sediment, yeast products, and other matter. Then they are PASTEURIZED to prevent contamination. As a consequence, there are NO live Yeast Cells in these beers. With the exception of such brewers as Coopers from South Australia and a number of small boutique beer brewers that are cropping up, all commercial beer if filtered and pasteurized.

This is an impossible feat to accomplish for home brewers unless he has specialized equipment. For secondary fermentation, you NEED live yeasts in your beer, in order to create the gas and frothy head, which is the reason you add the teaspoon of sugar when bottling. This process leaves a sediment of yeast at the bottom of the bottle, just as the first fermentation in the Fermenter left the slurry at the bottom after it was completed.

Filtering the beer has absolutely NO BENEFIT to the average home brewer unless they are creating what amounts to a commercial beer product, which means artificially 'gassing' your brew and/or they have purchased the specialist equipment to carry a secondary fermentation outside of the bottle.

Finitration, or the adding of FININGS, will settle excess sediment in your beer. Simply mix a teaspoon of Gelatine with some warm water, and stir into the brew after fermentation has ceased, and leave it for a day.

Incidentally, the tiny amount of yeast at the bottom of your bottle is sufficient to start off an entire new 22.5-litre WORT. I have used the sediment from Coopers Beer to great effect and achieved a truly excellent beer from it.

RECORD KEEPING

I have referred to keeping the BREWERS NOTEBOOK many times through this book. It is essential if you are to maintain a high standard because memory fails or make mistakes. Just as a policeman's notebook is sufficient evidence in a court of law, your notebook gives you sufficient evidence when arguing for a better brew.

Allow for a full page for every brew. It must be DATED, and I usually give each brew a number. By the time you get into the 500's you really know how much money you have saved. What you want to list on EACH PAGE and with EVERY BREW is the following.

- The INGREDIENTS you have used, or the make and type of the BEER PACK you have employed
- What type of HOPS used, and how they were boiled (If not hop essence/extract)
- The type of YEAST. If re-used, the bottling date and number of the brew it came from
- Any specific variation you used when setting up the brew
- The TEMPERATURE of the WORT prior to adding the Yeast
- The HYDROMETER READING (OG) of the WORT prior to adding the Yeast
- The SURROUNDING air temperature
- Whether you added HOT WATER or TAP WATER to arrive at your desired WORT temperature
- The DATE you TOPPED UP the FERMENTER if you did so
- Make a note ANY TIME you open the Fermenter Lid
- The SPECIFIC GRAVITY (FG) at Bottling Time
- The AMOUNT of WATER you added to the bottles if any
- Note if you are using RE-USED Crown Seals
- Note if there is ANY EVIDENCE of ANY INFECTION or DISTURBANCE in the brew, and describe it in detail, preferably with a photo attached
- Sterilizing method used for all equipment
- the heating or cooling method you used, if any, during the process
- Most Important: Write the DATE of BOTTLING on the CROWN SEAL with a marker pen. This is much better than writing a brew number and having to look it up to check its age. It also helps you keep track of multiple[le batches far more easily
- Put the DATE OF BOTTLING clearly in your NOTEBOOK, as it identifies the YEAST you have used and it is easy to match this to the date on your batch
- When you later DRINK this beer, write in your notebook things such as COLOUR and FLAVOUR, and all the ecstatic revelations that have come to you.
- Write in the calculated Alcohol Percentage

- Write in notes through the course of the brew regarding things like, how often bubbles were 'popping' and at what date you observed this
- Anything else you may wish to record, but nothing incriminating that the wife can stumble across.

It is absolutely frustrating to open a bottle in, say five or six months after bottling, and discover you have not one single clue what you did to make it so good. Without your NOTEBOOK you have no way to reproduce this golden elixir, or to find out what went wrong if it happens to be a dog. (almost impossible if you have correctly followed directions) You need to be able to trace what happened up to several years back and discover if you made mistakes, or need to get that same yeast again, etc.

If you intend to experiment, a thing I did a lot of, you MUST keep accurate records, or you will get lost in the maze of details. You are a SCIENTIST on the QUEST of BEER PERFECTION, so keep notes and make sure you know what happened to each recipe.

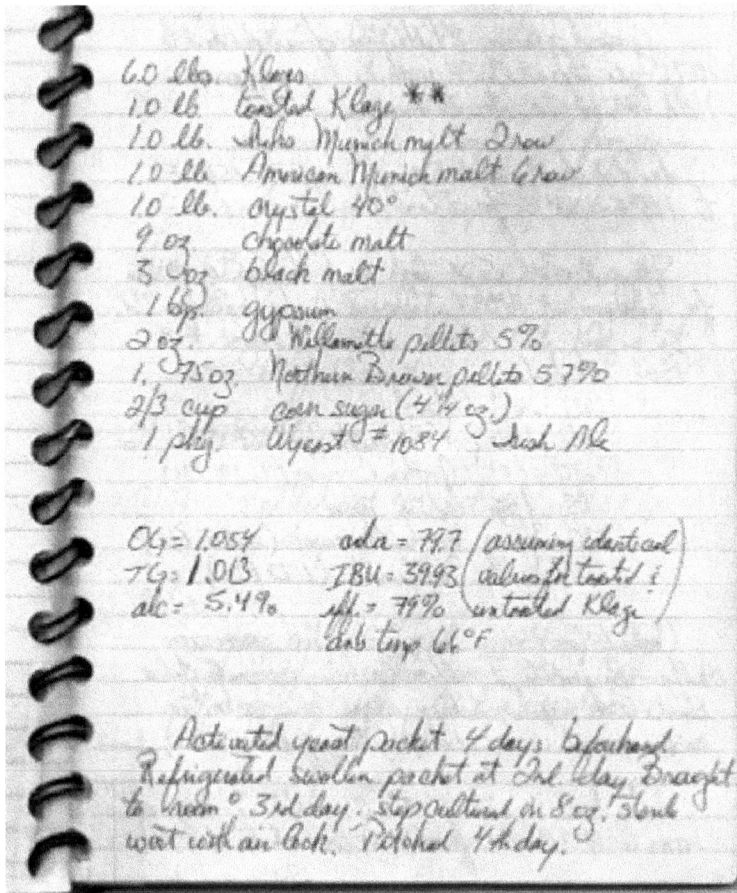

This is not Geoff's notebook, but it gives you the idea

EXPERIMENTING

Your first brew may be exactly to your taste, and all you ever want to do is to repeat this endlessly. I have no argument if this is what suits, but I can't say I have ever met anyone where this is the case. The real joy of Home Brew is the adventure and discovery we make in the quest for the perfect beer.

There is so much you can do to vary taste and appearance. More hops to make it a little more bitter, or more Glucose to sweeten it. Maybe it needs a little more MALT to add flavor? Or try DRIED MALT instead of MALT EXTRACT. You may even be so keen as to MALT your OWN BARLEY, which is a matter of sprouting the barley, drying it, and dry roasting it for 35 minutes in a hot wok. You may wish to experiment with GRAIN MASH and risk the odious Starch Haze it can create.

You may vary every single aspect to achieve a slightly different outcome, such as using a little HOP EXTRACT after you have boiled some HOP FLOWERS with some HOP PELLETS. Different types of sugar give slightly different outcomes, or use HONEY and try for MEAD, etc. There is NO FIXED RULE how to do ANYTHING, other than the basics of MALT, HOPS, YEAST, and WATER must be present in the correct balance in order to make beer.

You can even boil certain types of GRAIN, such as Crystal Malt, Light Malted Grain, Corn, Rye, Barley Flakes, Wheat Flakes, etc. You can do a MASH of various types of Grain. I have done all of the above, had a tremendously fun time doing it, and for the most part truly enjoyed the result.

As a small example, and excellent low-alcohol beer can be made with just 1KG of MALT EXTRACT, the usual amount of HOPS, WATER and YEAST and nothing else. You will be surprised with the clarity, appeal, and flavor. It is cheaper than buying 'pop' or soft drink, and it is a health-giving drink. That is the part of home brew many forget, it is GOOD for you.

Experiment away, but I can offer you a simple piece of advice, only change ONE thing per brew. Otherwise, because of the time delay in drinking the finished product, you will get lost in what exactly made the difference in the brew. Think of yourself as a SCIENTIST, keep ACCURATE RECORDS, and pay attention to the details, and it is a great adventure.

Be sure to write everything down in your BREWERS NOTEBOOK and do NOT allow to leave things and hope you will remember them. There are TOO MANY DETAILS, and you simply won't recall them all. Just write down everything, and you will be grateful down the track that you did.

"I am a firm believer in the people. If given the truth, they can be depended upon to meet any national crisis. The great point is to bring them the real facts, and beer."
- Abraham Lincoln

COCKROACH in the WORT

I saw it floating there, and was frankly shocked. I took so many precautions before and after fermenting the WORT and I have absolutely no idea how it got in there, other than perhaps it had flown onto the underside of the lid before I sealed the Fermenter.

I presumed the entire brew would be ruined, but in my research, I was surprised to learn that Cockroaches are extremely hygienic creatures. If you touch them, they run away to CLEAN themselves. It is not something you expect, but rather than toss the lot, I let it run to see what the result would be.

Fermentation went along as usual, in fact, perhaps a little better than usual. I obviously had plucked out the cockroach when TOPPING UP and then thought that I should just bottle the beer anyway because the alcohol would have killed the germs, if any. The beer looked fine, smelled fine, seemed to be all good, so I bottled as usual and marked the bottles with a "C" for Cockroach.

Well, come taste testing time I was extremely tentative. However, the real surprise was not that the beer was OK, but that it was absolutely excellent! It was a great tasting brew, one of my best. I started to wonder, and in the end, I realized that the adding of a little PROTEIN to the WORT can help with flavor. No, I didn't add any more cockroaches, nor even the odd fly, I started to add a small amount of GELATINE to the mix.

This is the protein power we using for Fining the WORT. (reducing cloudiness, which is sediment in suspension) Just the tip of a teaspoon seemed to really help the flavor of the beer. Many years later, my son *(Editors Note: ME ... The Editor of this book)* was giving a lift to the son of an English Brewer, who said he had to get back to England to help his father with the yearly brew of Hogs Head Beer that their local pub was famous for. It was a secret recipe known only to the family for generations.

Apparently, because of my telling him about the Cockroach in the WORT, my son said, "I reckon I know what the secret part of the recipe might be."

The other lad laughed and asked him to guess. My son said, "You throw a Hogs Head in the WORT."

The brewer's son stopped laughing and said he must not tell anyone. Well, the secret is out, a little PROTEIN in the WORT can assist the flavor. Some add blood, others add a little grain, there are many things you can do to modify and improve the overall flavor of your beer.

*(**Editors Note**: You can imagine how much I laughed when the Harry Potter books came out, and the favorite tipple there was Hogs Head Beer!)*

Milk is for babies. When you grow up, you have to drink Beer.
Arnold Schwarzenegger

BEER as a FIRE EXTINGUISHER

S ince your FERMENTER may be kept in a very small space, as one may well have to do in some circumstances which require the WORT to be kept warm, for instance, or where accommodation is tight, I thought I had better ring the local FIRE DEPARTMENT to check if there were any specific hazards I needed to be aware of.

The good officer I spoke to advised me that there was no danger whatsoever and pointed out that the Carbon Dioxide the Fermenter emits was a FIRE SUPPRESSANT. Your brew is effectively a FIRE EXTINGUISHER! Furthermore, he added, a bottle of beer, opened and shaken up creates a CO_2 emission unit that also gives forth a fire suppressing foam, and that it makes a convenient FIRE EXTINGUISHER in an emergency. It cannot be used for fat, oil, petroleum or similar liquid fires as the 'squirting force' of the fluid ejected can actively spread such a fire. He also stressed to NOT use your BEER FIRE EXTINGUISHER on ELECTRICAL FIRES as the fluid can form an electrical circuit that can come back at you. We do not want to eliminate a promising brewer such as yourself!

I have long considered putting out my credentials as a FIRE RESISTANCE OFFICER when queried about my desire to drink beer. "I am not drinking Madam, I am mere testing the fire resistance of this fluid". As they have long said, a good brew certainly puts out a fire inside you!

This extremely helpful officer then mentioned it was safe in motor vehicle fires. In the case of the motor catching fire, spray the beer over it, and shut the bonnet, to exclude air. To be truthful, a bottle or can of soft drink works in a similar fashion, but it just doesn't have the same ring to it as a BEER FIRE EXTINGUISHER, I am sure you will agree.

However, in most cases, it is probably better to just call the fire department and drink the beer while waiting. Better still, have a PROPER fire extinguisher on hand, and drink your beer in comfort knowing you have been a good boy scout

Give me a woman who loves beer, and I will conquer the world!
Kaiser Wilhelm

How to GET RID of BOTTLE SEDIMENT when TRAVELING.

Many people like to go camping and take along a few cartons of home brew to celebrate getting to the other end. Obviously, taking home brew in a car means it gets shaken up, and the sediment can be stirred up in the bottle. If you don't have days for it to settle at the other end, you need to know a way to make your beer drinkable again. (As a note, if you do a 'travel brew' add extra sugar to make more gas, which also helps suppress sediment movement when traveling)

I came across the obvious the hard way, taking a couple of cartons in the back while making one of the regular camping and hunting trips my son and nephew would make 'Out West". Furry beer full of sediment is not the most pleasurable thing to drink, but of course, waiting for a day or two before it settled was out of the question. But I started to think on how best to avoid this.

If your beer forms up a proper 'matt' at the bottom, this is less likely to be disturbed when being bumped long and can travel quite well in a car. If it is sludgy, it will not. If you wish to carry on, and still take this beer with you on your next trip, this is what you do.

Decide how many bottles you wish to take and put them into the refrigerator for as many days possible before departure. Get them AS COLD AS POSSIBLE, without freezing. This assists in 'fixing' the gas in the liquid. Have ready sufficient clean bottle and simply, and carefully, DECANT each bottle into the fresh ones, and seal it right away. Have the receiving bottle at an angle sufficient to allow the efficient use of its funnel, and lessen the loss of fizz. You lose a little gas, but you also lose the sediment. (Hence the suggestion of putting away a 'travel batch' that has more gas than usual)

Your resultant new bottles are free of sediment and can travel freely without concern. Of course, they will not 'mature' anymore, but the point is you are taking them to drink, not gaze out the window wondering how they can improve themselves.

The FIRST beer, the one you are decanting FROM can be a little frothy when extremely cold. Simply wet the insides of the bottles you are decanting TO, and it will reduce the froth.

The TYPES of ALCOHOL

There are many types of alcohol, but the kind we are interested in is that found in the usual alcoholic beverage. C2H2OH is ETHYL ALCOHOL or Ethanol, and it is made initially be the fermentation of starches converted to sugar, or of other sugars, by yeast.

The HIGHEST alcohol content that can be obtained by fermentation is around 14%. Any higher level of alcohol must be obtained by distilling the yeast-fermented liquor. This is done by heating the liquor and using a 'still' to catch the evaporation. Alcohol 'boils' at a lower temperature than water (78.5 C or 173.3 F) and so as a consequence, when the liquor is heated to that point above the alcohol boiling point, but below the water boiling point, the alcohol is vaporized. It can then be collected in the still and converted back to a liquid.

This is how all alcoholic spirits are made. This first distillation can be further concentrated, up to almost 100% pure. In this concentration, it is an irritant and a poison, even in moderate amounts. Remember, this whole process starts with Yeast converting SUGAR into alcohol and CO_2.

Any other type of alcohol, such as methyl or wood alcohol is a violent poison and must never be drunk. Blindness and death can be the result. *(Editors Note: I once spoke with an alcoholic who had moved onto methylated spirits. I asked if he understands what would happen, and he answered that he did, but there was a point when normal alcohol no longer did it for you and that methylated spirits did. He knew it would kill him, but he no longer cared.)*

One of the turning points in my life was when, under the influence of prescribed medication, I suddenly thought it would be a good idea to drink White Spirits. I survived, but realised that natural medicines, such as beer, were far better for you than the psycho-active pills handed out for depression and anxiety. Nutrition and common sense will help you survive better than any psychiatrist.

Alcohol is usually expressed as percentage by volume. Most commercial beers (except the lights) re around 5% Alc/Vol. This means that one-twentieth of the contents are pure alcohol. Table wines around from 8% to 14.5% Alc/Vol. As a note, when a wine is at 14% it generally means it has no preservatives and is often a better quality product. `Fortified wines, such as port and sherry, are around 18%. (Fortified means that alcohol or brandy has been added) Spirits range from 18% up to 45% Alc/Vol.

The term "proof" changes, with a different reading for the same alcohol content being given, from country to country. We go over this in an earlier chapter, "Note on World Proof Systems".

BRITISH - 100% pure alcohol is 175 Proof

U.S.A. - 100% pure alcohol is 200 Proof

Metric (G.L.) n- 100% pure alcohol is 100 Proof

Blood Alcohol Level in DRIVERS

Much is made of the importance of your blood alcohol level when driving today. I will not dispute the wisdom absolutely generalized limits of this, but the fact remains, different people can drink the same amount of alcohol and get vastly different readings.

Assuming that an average sized person had consumed no significant amount of alcohol for an hour, and if that person drank quickly on an empty stomach, they would be over the 0.05% blood alcohol limit of most countries and considered unsafe for driving.

This represents 82 ml (2.7 fl oz) of 45% spirits, 204 ml (6.9 fl oz) of fortified wine, 739 ml (25 f oz) of average strength beer or 1680 ml (57 f oz) of 2.2% light beer. In

"No more for him - he has to drive."

practice, however, the body starts to metabolize alcohol as soon as you start drinking.

Remember we are talking a percentage by volume of alcohol in the blood. Logically, and in practice, a large person who has a great volume of blood can drink more alcohol and remain at the same blood alcohol limit. This is partly why women appear to often get drunk quicker than males, as most men simply have more blood with which to dilute the alcohol.

Size is not everything, the condition of the liver, prescribed medications, the degree of fatigue, a person's emotional state and the level of oxygen in your system can ALSO affect the BAC (Blood alcohol concentration). The message is simple if you are drinking more than the hourly volume of the alcohol described above, you are at risk of being charged for drink driving.

This is another reason why I so often brewed my ales to have a medium to low alcohol content. I can drink more, and not feel at risk when hopping into the car. The taste and flavor of the beer is not impacted if you use MALT instead of SUGAR, as one example, and it gives a lower alcohol reading at the end.

Many countries have a 0.08% alcohol limit, which is far more sensible. If so, simply adjust the above quantities by dividing by FIVE and multiplying by EIGHT. In other words, you can drink 1230 ml (41.6 fl oz) of a 5% beer in an hour before hitting 0.08% BAC.

All the above is based on 7.39 ml or one-quarter of a US Fluid Ounce of pure alcohol (100% Alc/Vol) raising your blood alcohol by 0.01%

Effects of ALCOHOL in the BODY

Alcohol is rapidly and almost completely absorbed by the body and requires no digestion. This does not apply to the various things it may be mixed with. It is a quick source of energy. If sufficient alcohol is taken the body can access approximately one/fifth of its energy needs this way. If more carbohydrates and fats are eaten than the body needs, it will use the alcohol first, and store the excess of the others as fat.

Alcohol is NOT a stimulant, as commonly thought. It is a SEDATIVE which depressed the central nervous system in much the same manner as tranquillizers. In sufficient quantity, it reduces inhibitions and people will do and say strange things under the influence of booze. A large, concentrated dose (such as a full bottle of spirits) drunk quickly can cause a person to fall into a coma, and even trigger death. He drank himself to death, is a saying most are familiar with, but the truth is, he tried to cover up his issues with enough alcohol that it killed him.

In moderate amounts, alcohol seems to reduce anxieties, worries and physical pain. Somehow they always seem to reappear the next day, however. Doctors say that moderate use can reduce blood pressure, but if you have high blood pressure, it is far better to see your doctor than to just drink and hope it goes away. I know I tried, often. A small quantity can put you to sleep, a large amount can make you restless and unable to sleep.

The simple fact is, it is a poison that in small doses has little by way of harmful effects, but severe dehydration and hangovers can be the result of excess consumption. It can place a heavy load on the liver, and when someone is a constant drinker and neglects nutrition, a fatty liver can develop. Cirrhosis of the liver can occur with prolonged, heavy use. Excessive drinking combined with a poor diet can result in a mental disturbance known as Korsakov's Syndrome. NB: Good nutrition means quality food, not lots of fast food

An average sized person can metabolize (process) about 40 ml of 37% spirits per hour or around 300 ml of 5% Beer per hour. Obviously, if your alcohol intake per hour is higher than this, your blood alcohol concentration (BAC) will rise accordingly. There is no 'average' person and everyone will vary in how effectively they can absorb alcohol, but I can assure you, sucking on mints before a breathalyzer test does NOT reduce your reading. Your emotional state can, and your blood oxygen level can, but if you drink enough, nothing stops the police from charging you with Drink Driving.

In the US the more civilized test of seeing if you can walk a straight line is, despite the critics, a more accurate measure of how alcohol is AFFECTING you. Touching your nose with your fingers test hand to eye coordination, a thing that a very drunk person loses, so another valid test.

Around 20% of the alcohol consumed it absorbed through the stomach wall, and most of the rest is absorbed in the upper small intestine. Around 90% of what you consume has to be metabolized by the liver. Food in the stomach, in particular, fatty food, can delay but not stop the absorption of alcohol into the blood. Coffee has ABSOLUTELY NO EFFECT whatsoever in getting rid of alcohol out of your system. It may wake you up a bit, but it will not sober you up.

If a person sticks to mid to low strength beers and has a good diet, it is very unlikely Mr. Korsakov will knock on your door with his syndrome.

Editors Note: It is worthy of note that Geoff went to 91 years with full mental acuity. He was laughing and joking with full awareness of his surroundings until the day he passed on. Despite the fact he drank often, he always maintained a very healthy diet and a decent level of exercise

Editors Note: Korsakov's Syndrome is otherwise called 'alcoholic dementia'. I had an aunt who suffered this, and it is not a pleasant thing to watch a loved one go through. More information: https://www.alz.org/alzheimers-dementia/what-is-dementia/types-of-dementia/korsakoff-syndrome

Geoff walking near his home on the Gold Coast, Australia with his son, Michael (the editor of this book) Taken two years before he passed on.

Acidity / Alkalinity and the pH Scale

The Chinese speak of Yin/Yang and the importance of balance between the characteristics of life. In brewing terms, the "Yin/Yang" is the acid/alkaline balance. The balance of acidity is extremely important, critical in some circumstances, in producing a fine ale.

Town water, which most will use in their brewing, is slightly alkaline. Beer needs to be slightly ACID and many brewers will add a teaspoon of citric acid or ascorbic acid to the wort prior to pitching the yeast. However, this practice is not necessary, as the working of the YEAST upon the WORT of itself produces a mildly acidic solution, with the resultant beer that starts with town water ending up with a pH reading between five and six. (Seven is neutral)

If you wish to add acid, be sure you are using ascorbic acid, because most health food shops will give you sodium ascorbate or calcium ascorbate when you ask for "Vitamin C" powder. They are all concentrated Vitamin "C" but they are not ACIDIC to anything like the level of Ascorbic Acid.

It is important to test your water to see how acid or alkaline it might be. This is easy to do, and there are test strips for this purpose at every pool supplies shop. If your water is more the eight in alkalinity, you will need to add a teaspoon of citric or ascorbic acid. RAINWATER, surprising to most, is slightly ACID if there is no significant air pollution in your area. (Around pH 6) If there is a lot of smoke and smog, it will be even more acidic.

Slightly alkaline is not a problem, but too acidic or too alkaline is and needs to be addressed prior to adding the yeast. A very small amount of bi-carb soda will reduce acid. You need to test and adjust according to your local area needs. Distilled water is NOT suitable for brewing. Rainwater is not necessarily better than tap water, or expensive spring water, but as long as the acidity level is acceptable, anything will suffice, but NOT distilled water. It is dead water.

The pH Scale ranges from 1 to 14, 7 being neutral, 1 being extremely acid, and 14 extremely alkaline. You want your WORT to be around the SEVEN, and a little alkaline is fine because, as mentioned, the yeast itself creates a mild acidic reaction in fermentation.

To ascertain the level of any liquid you use LITMUS PAPER. (Remember the term 'Litmus test'? This is it!) This paper will change color according to the acid/alkaline balance of the fluids you are testing. When asking for litmus paper, you want something that will change its color inside 30 seconds. You also want to buy a ROLL, instead of expensive strips, and you only need an inch of litmus to test any liquid. Simply dip it in the fluid, and it will quickly give you a determination.

Generally, each dispenser has a 'color chart' on the side, which tells you the level of acidity/alkalinity present. You do not need expensive litmus paper, such as Clinistix Litmus, which is used in laboratory conditions. You only need a general level of the acid-alkaline balance.

Our blood is generally alkaline at 7.4. Life is only possible when the blood lies between 7 and 7.8 alkalinity. Beer needs to be within 5 to 6.4 Acid.

STARTER BOTTLES

Some brewers like to have their YEAST actively working BEFORE they introduce it to the WORT. The way you do this is with a STARTER BOTTLE. It is simply a glass bottle or jar into which you put the YEAST you intend to use in your brew, along with a little MALT, SUGAR, and WATER. You can use some of your WORT if you wish. The idea is that it allows the yeast to get started with the fermentation process, which means it will be growing and multiplying.

It is a very useful thing to do if you want a quicker start to your brew in very cold conditions, and where you are using bottom fermenting yeast. This process is also used by brewers who mash their own grains, when it has taken time to COOL the WORT to yeast pitching temperature, and where infection can set in if fermentation does not start quickly. It can be all of the above in one, as well.

Other than this, it is an unnecessary procedure for the average home brewer in most conditions. It is simply easier and no less effective to just throw the yeast into the WORT directly in the FERMENTER. However, if you are experimenting with various types of YEAST, Starter Bottles are essential for a good result.

The simple way to make a Starter Bottle is to use something like a milk bottle. You will need a cork bung to suit the bottle, and you drill a hole into it to insert your AIRLOCK, which obviously has to create an airtight seal with the bottle.

You can also use a screw lock bottle, and punch a hole in the lid, using a rubber grommet to create the seal with your AIRLOCK. The ideal size if around 600 ml (one pint). Always sterilize the jar and clear glass is the best option, as you want to be able to see how things are going. Store in a dark place. You can even use a jar with plastic wrap across the top, sealed with a rubber band, with a small hole spike in it to allow the gas to escape.

NEVER put YEAST, MALT SUGARS, STARCHES and WATER into a tightly sealed jar, it WILL BURST.

Another use for your STARTER BOTTLE is to add some of the WORT to it AFTER fermentation has commenced. This way you can see how your mini-beer is brewing. As with the fermentation process, you will see millions of tiny bubbles rising. I once used a milk bottle with a balloon stretched over the lip. It grew every day till it burst.

Fermentation is generally complete after a few days, the bubbles cease, and you see a whitish yeast settled on the bottom of the bottle. This small amount of yeast is sufficient t start off a new batch. Yeast will continue to multiply as long as it has sugars and nutrients to feed it.

You can't be a real country unless you have a beer and an airline. It helps if you have some kind of a football team, or some nuclear weapons, but at the very least you need a beer.
Frank Zappa

Calculating Approximate Alcohol Levels based on SUGARS USED

The amount of FERMENTABLE sugar in the WORT determines the level of alcohol that will be reached during fermentation. There is already SUGAR in the MALT, or BEER PACK, you are using. There may also be sugars from mashed grains you are experimenting with. What we are talking about here is purely the SUGAR YOU ADD and how each type converts to Alcohol.

SUCROSE, or cane sugar, coverts half it's weight into alcohol by the time fermentation is done. The remainder is released as Carbon Dioxide Gas. Ordinary white cane sugar is the most efficient sugar for brewing purposes. We will take this as the standard, and use this as the comparison by which we give percentage conversion rates for other sugars. These figures are an accurate approximation but are not exact. There are MANY variables in Malt Extracts alone, but it is a good rule of thumb.

White sugar converts 100% during the fermentation process. In comparison, the following sugars convert to the percentage listed:

TABLE C

Raw Sugar	99%
Brown Sugar	97%
Invert Sugar	80%
Dried Malt Extract	74%
Liquid Malt Extract	66%
Dextrose Powder	83%
Honey	80%
Glucose	50% (Varies with quality from 40% to 67%)
Light Malted Grain	60% (Varies with MASHING technique)
Other Grains	60% (Must be MASHED to convert Starch to Sugar, otherwise Nil %)
Crystal Malt	37% (When BOILED in the WORT. Does not need mashing)
Roasted Barley	6% (Used mainly in STOUT to flavor and color)

Now, to covert this to an alcohol value, we have to go back to our basic method of calculating alcohol, but adjusting for the type of sugar we have used. Remember, the OG and the FG are now based on a sugar that has not fermented properly, so we need to adjust this when using the various options.

EG: If we have an OG of 1.040 and an FG of 1.010 in a WORT that had 500gm of HONEY, we have to adjust for the fact the honey component produces 20% LESS alcohol than white sugar. So one way we can calculate this is as if you had used 400gm of SUGAR. But when changing formulas, adding malt and a variety of sugars, it gets a lot more complicated.

As we have done all of our calculations using a 22.5-litre WORT (5.9 US Gallons) we derive a table that gives a value of Alcohol to Sugar. I will give this in 1% alcohol increments. (Next Page)

TABLE D

Grams	US Oz	Alc/Vol
432	15.2	1%
864	30.4	2%
1296	45.7	3%
1728	60.9	4%
2160	76.1	5%
2592	91.4	6%
3024	106.6	7%

Editors Note: I have gone through the extensive list of letters to Universities and brewing companies, as well as product manufacturers, that my father wrote to requesting detail information for this specific section of his book. I can attest to his diligence in getting this area correct.

In the below example, you have used 1k Liquid Malt Extract, 1k Dextrose Powder and 180gm priming sugar in your bottle.

1000 gm Malt	660 gm Sugar Equivalent	
1000 gm Dextrose	830 gm Sugar Equivalent	
180 gm Priming Sugar	180 gm Sugar Equivalent	
Overall	660 + 830 + 180 = 1670 gm	= 3.8 % Alc/Vol

You are working with an EQUIVALENT SUGAR CONTENT of 1670 gm which gives you just under 4% alcohol as the expected concentration of this brew.

In a further example, we have a TARGET to aim for and need to work out how much of each ingredient we need to reach it. In this case, you wish to mash your own light malted grain for an ALL MALT brew. You will be using NO WHITE SUGAR except for priming the bottles at bottling and you are aiming to get a 5% alcohol Beer. You look up 5% alcohol, and it corresponds to 2160 gm (76.1 fl oz) of SUGAR. But remember that we ALSO have to subtract the 0.5% of alcohol that comes from 150 gm of the PRIMING SUGAR.

So our start figure is 2160 - 150 = 2010 gm. Your LIGHT MALTED GRAIN has an equivalent ratio of 60% to white sugar, so how do we work a reverse percentage? It is **not** 60% of 2010 gm, 2010 gm is 60% of what we need if we are using MALT. So we MULTIPLY 2010 by 100 and divide by 60 which equals 3350. This means 3350 gm (118.1 o) of LIGHT MALTED GRAIN MASH must be used to get to an overall 5% alcohol in the bottle.

It can get very complicated the more variables you use, but if you want to know a final likely alcohol content when you are using a variety of sugar types, this is the only effective way to do it.

INVERT SUGAR

nvert Sugar is ordinary white sugar (Sucrose) which is BOILED with water and ACID until its color changes from clear into a very pale gold of yellow. The SUCROSE has not been changed into its two component sugars, DEXTROSE and FRUCTOSE (fruit sugar) in approximately equal quantities.

The enzyme INVERTASE is YEAST performs the same function during the fermentation process, so it really is not necessary to use Invert Sugar for brewing. Some brewers will tell you that inverting the sugar will lead to quicker fermentation and a better flavor. I have used it on occasions to test the theory, but my taste buds discern no significant difference. However, you may wish to try it, you this is how it is done.

Dissolve one kilo (2lb 2oz) of sugar into 750 ml (25 fl oz) of boiling water. Add one LEVEL teaspoon of citric acid and stir in. Bring the water to a vigorous boil whilst stirring. Once it boils, you have no need to keep stirring. Lower the heat to a SIMMER for about 25 to 30 minutes. At this time you should start to see the clear liquid change to yellow. It is now ready for use.

These measurements make exactly 1 liter of invert sugar, so if your recipe calls for 300 gm of granulated sugar, use 300 ml of the Invert Sugar instead. Taste it when you have finished. It is quite delicious and I am sure a good cook could find other uses for it in food preparation.

There is an old wives tale that drinking FRUCTOSE after alcohol will lower your blood alcohol level, but I suspect this just comes from drunk farmers eating some fruit and feeling like they are sobering up. Peaches and cheese do go well with beer, though. No one can deny it.

The Thermometer

The most common Thermometer for home brewing has the customary glass tube which is affixed to a steel frame. The frame has the graduation marks on it. There are other types, but this variety is inexpensive, strong and is, of paramount importance, easy to clean and sterilize.

Your ordinary Thermometer, used to air temperature, is generally on a wooden frame. Wood is absorbent and is NOT SUITABLE for general brewing duties. However, as with everything, if it is all you have, as long as you sterilize it, it can be used until you find something more suitable.

The one I use I have had for many years and is still calibrated in Fahrenheit.

Most used Thermometer nowadays. Easy to clean and sterilize, and it can measure temperature at different levels of the Wort

It is marked between 70 and 80 degrees F (21 - 27 C) with two lines and between these are the words "Add Yeast". This is a misleading instruction. It is all well and good in most situations, but if it is very cold you may want the WORT to be hotter before adding yeast, to ensure a good start to Fermentation. Further, if you are using a TOP FERMENTING YEAST, the temperature should not be allowed to drop BELOW 75F (24C)

A Thermometer is necessary to calculate the alcohol strength with your Hydrometer. The Hydrometer is calculated to give the correct reading at 60F (15.5C) and as the WORT often varies from this, the Thermometer gives you the reading you need to adjust for this. You can easily add 0.5% of alcohol to your final bottle by not using a Thermometer with the Hydrometer. As one example, if the WORT reads 90F (32C) and you do NOT use the Thermometer to correct your projected alcohol, relying on the Hydrometer alone will put your estimation 0.5% lower than it will actually be.

Obviously, your Beer Thermometer is used for measuring the temperature of liquids, but it shows air temp when left out. If you ever get involved in mashing your own grains, temperatures of 180F (82C) will be used, thus ruling out your household Thermometer. If you are serious about your home brewing, get the right equipment.

Editors Note: You can buy Hydrometer / Thermometer as a package on Ebay for $12.50. There are also heating pads and temperature strips to control the heat of the WORT.

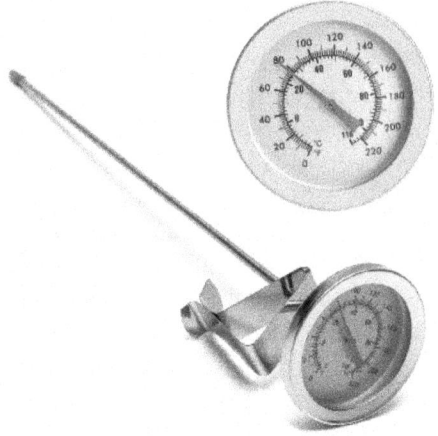

OPEN FERMENTING

This term 'Open Fermenter' is exactly what it suggests, you have NO LID on your Fermenter. After many years of experience with closed Fermenters, and knowing how important this was to have scrupulous cleanliness and sterilization of all equipment, I was reluctant to even try this centuries-old method. Fresh air, usually so necessary to health, is the enemy of home brew.

Air carries air-borne bacteria and wild yeasts, as well as other organisms harmful to your WORT. The mixture of warmth, sugar, and nutrient is just about the perfect home for any number of beasties that would just love to get in and grow there. Unless these blighters can be kept OUT, your brew stands little chance of success. That said, even though the unhygienic and air-filled WORTS of so many home brewers turns out what I call a bad batch, these people drink it happily, saying it is terrific. This is stuff I would spit out, but each to their own. Needless to say, however, you can easily understand the trepidation with which I approached open top Fermenters.

The difference and the reason Open Top Fermentation works is because of the YEAST you use. Top Fermenting Yeast gets to work and creates a blanket of CO_2 gas over the top of the brew. It is a heavy gas, and as long as the Fermenter is out of wind, the gas just sits there and forms a protective covering. After it rises from the WORT, it just sits there. Now, while we say "open" Fermenter, you still cover it with a towel to provide protection from flies and insects. The thing is, as soon as fermentation is complete, the CO_2 gas no longer is being evolved and no longer forms a band of protection. You must bottle Open Top Fermenters AS SOON AS FERMENTATION IS COMPLETE.

Because you cannot let it sit for several days, and have the sediment drop to the bottom, many people use FININGS, such as a teaspoon of Gelatine stirred in, to precipitate the sediment more quickly. But as this whole thing is such a risky process overall, you might suggest, that there has to be some sort of great benefit to using a Top Fermenter Yeast.

The MAIN benefit is that Top Fermenting Yeasts work much better in HOT CLIMATES. Where you read of beer being produced in Ancient Egypt, you can be assured they used Top Fermenting Yeasts to achieve this. Obviously, you have to get your fermentation underway as soon as possible, so if trying Top Fermentation, use a STARTER BOTTLE as outlined a few chapters back. Even so, using the Top Fermentation Method, the beer is clearer at the completion of fermentation and the beer matures more quickly in the bottle, so can be drunk much sooner.

Therefore, you don't need to hold large stocks of beer sitting in the shade, maturing for months. Things like this can be very important in hot climates, and when you think about it, the whole concept of beer came out of Mesopotamia and Ancient Egypt, all hot climates.

The main DISADVANTAGE is that Open Top Fermenting MUST BE DONE in a warm climate, or in circumstances where constant warmth can be provided. Your WORT temperature should not fall below 77F (25C) and you must absolutely be ready to bottle as soon as Fermentation is complete. You have a maximum of 24 hours to get the brewed beer into bottles. And obviously, during this phase, there is a significant risk of infection, much higher than a closed fermenter.

What sort of vessel should be used for Open Fermenting? Well, your ordinary Fermenter is fine, just leave the lid off and throw a towel over it. It has the benefit of already having a tap fitted, which is essential unless your siphoning abilities are superhuman. So you already have the tools needed to give this a go.

However, many people just used plastic garbage bins, clean ones, of course. The main thing is that the Fermenter should not be made from PVC (Polyvinylchloride) as this can become poisonous under certain conditions. One of the best things to use is one of the water bottle containers used in office supplies. These are made from polyethylene ter phthalate (pet) and are very safe to use. Most are 20 liters, however, so you need to adjust for this. If your container is METAL, it should be enamel or stainless steel. Aluminum and Cast Iron are porous and can affect the flavor of your brew.

You can buy a tap from any home brew shop, and fit this into your chosen container. The tap should be 50 ml (2 inches) above the base of your container, to allow for sediment to drop below it. If you cannot muster up a tap you will need to SIPHON everything out for bottling. Not recommended, as it is too easy to stir things up, but it works.

If using a SYPHON, it is best to tailor fit yours to your container, so that it goes to a very specific length that stops 50ml (2 inches) above the base of the container. You can create a wire bracket to hold it in place, and it is significantly better if you fit two lengths of hose to a glass 'u' tube, one end to go into the container, the other to have the longer hose sitting outside. Really, it is easier to just buy and fit a cheap tap.

Cleaning and sterilizing must be as scrupulous with Top Fermenting Brews as with everything else you do. Never say "It looks clean" and use something, this is a guaranteed way to lose a batch. Tea towels or muslin that you use to cover the vat must ALSO be clean and sterile. Just a rinse in bleach is sufficient.

It is very hard to advise any beginner to start with Open Fermenting. You really need a little experience to understand what happened if something goes wrong. However, it is quicker, does not need specialist equipment, and as long as you have read and understood everything, as well as accessed the proper YEAST, there is no reason not to give it a go.

For your FIRST brew, I recommend forgetting about recipes and using a BEER PACK designed for Open Fermenting. Coopers have these in Australia, but wherever you are, there will be something you can use. I have used the COOPERS brand packs and found them to be more than excellent, and their YEAST is suitable for Open Fermenting, Further, I have found I can make 40 bottles from one can of their concentrate, an extra ten to what they advise, and have noted little difference in flavor or taste.

As the years rolled past, and I have literally thousands of brews under my belt, with hundreds of experiments that went spectacularly right or equally wrong I

BEER O'CLOCK

found the convenience of the BEER PACKS to be hard to go past, and this is specifically so when it comes to Open Fermenting.

The DIFFICULTIES of Open Fermenting are no what you might expect. Because you have no AIRLOCK, you have no immediate and clear measure of when the brew has finished the fermentation cycle. And remember, leave it in there too long, and it WILL go off. You have an open top, so you have to check it visually. Just lift a corner of the covering material and you should see a think, brown and white frothy scum on the surface of the WORT. This is EXCELLENT, and what you want.

In three to four days, this will subside, and when the surface is CLEAR and FREE of Froth and Bubbles, or almost so, fermentation is complete. Check with your Hydrometer, using a test jar you decant a little of the WORT into. (Do NOT put the Hydrometer directly into an Open Top Fermenter) If you get a reading of around 1.006 (corrected for temperature) the beer is ready to be bottled inside the next 24 hours.

NOTE: In HOT weather, you can be ready to bottle in just two days. If it takes more than 4 days, the temperature of your WORT has probably dropped to low. Keep it to 77F (25C) Further, do not TOP UP an Open Top Fermenter. You can add sufficient water at the START of the process, thus removing entirely the need for TOPPING UP completely. (There is no lid for the froth to overflow onto)

Some brewers like to 'skim the froth' that appears on the top of the WORT in an Open Fermenter, and I can only stress that this is foolish. It is a layer of protection and should be left in place. Skimming can only increase the possibility of infection getting into your WORT. As a note, I have carefully inspected the amount of detritus and yeast left at the bottom of the Fermenter after both Top and Bottom fermenting yeasts have been used. The end result is virtually identical in either case.

Now, despite the fact I recommend against skimming, it really comes down to the type of YEAST you are using. If the frothy layer sinks to the bottom of the WORT after a couple of days, leave it alone. If it is the type that stays on the surface forever, you may wish to skim it off.

REMEMBER, you CANNOT leave the WORT in an Open Fermenter once the fermentation process is complete. I recommend you have EVERYTHING PREPARED for bottling WELL BEFORE you need to decant your Wort into the bottles.

If you cannot leave the Fermenter INDOORS or UNDERCOVER and AWAY from the wind, do not use Open Fermentation. That said, I have brewed up Coopers Stout using the open Fermenter method, and found I have been able to drink it just SIX DAYS out from bottling. Most Open Fermenter beers can be drunk within two weeks of bottling. It is the strongest argument for the process.

SUMMARY: The *Open Top Fermentation Process* works very well if you want to build up a stock of beer quickly. You can be bottling a new batch every four to five days, and drinking it two weeks later, as opposed nine days fermentation in a closed Fermenter, and waiting up to two months before drinking.

The ABSOLUTE IMPORTANCE of STERILIZATION

This is the MOST IMPORTANT chapter in the entire book. If you do not do this right, everything else will be for naught. Sterilization is the most important aspect of home brewing because you are working with "Bug Heaven" in a Wort, and ANY contamination will grow and take over all your good work.

Failure to sterilize all equipment and/or allowing air to be exposed to the WORT after fermentation is done without question are the two main causes of failures with Home Brewing. Neglect in either of these two areas will result in batch after batch of bad beer.

Cleanliness goes without saying, as anyone who goes to the trouble to sterilize their equipment will without doubt also clean it first. We have covered the 'how to' on this subject earlier in the book but I want to close the technical section by stressing and emphasizing yet again the imperative that is sterilization.

Closed Fermentation means the top to the Fermenter must only be removed for Topping Up and to check immediately prior to bottling. You must strive to keep air and the outside elements away from your precious WORT. Leave the top off and it almost ensures infection will get in.

Ordinary Household Bleach is all you need to keep all your equipment sterilize. Cheap and effective, you simple need to rinse the equipment well before use.

Open Fermentation has the carbon dioxide forming a protective blanket over your WORT, and the top dwelling YEASTS form another barrier. But when fermentation has ceased the brew is vulnerable. You must bottle quickly or it WILL become infected. Not perhaps, it WILL become infected by something.

Failure in either Sterilization or protecting the WORT from bugs is 99% of all troubles the home brewer will suffer. But if you follow the rules do as this book suggests, and strictly adhere to them, you will be rewarded by a never-ending supply of cheap, good beer. If you have been a brewer who has suffered a number of bad batches, the reason is almost certain to be found in these simple principles. Sterilize the equipment and protect the WORT.

"Always do sober what you said you would do drunk. It will teach you to keep your mouth shut."
Ernest Hemingway.

COMMERCIAL BEER MAKING

I did set up a series of vats for commercial beer making as I wanted to do a test case and report on it for this book. This entailed mashing grains, using filtration and gassing the bottles at the end. I can only report limited success, largely due to the constraints of the budget.

However, there is a story I would like to close with, detailing my discussions with the head of Castlemaine Perkins Brewers here in Australia. During my days as a Real Estate Agent, I met many people and chatted over a vast variety of subjects, but being an avid home brewer the chance to glean some information from the man who ran one of the top breweries in Australia was too great an opportunity to pass up.

I asked a lot of technical questions and was impressed by how articulate the man was, and the depth of knowledge he had with every single aspect of the brewing process. He too stressed how essential sterilization was, the type of yeast, the length of time in the vat, etc. All just echoed what I have said here, but the thing that really got me was that I had noticed, despite following all procedures to the letter, one in ten brews was just not as good as the others.

When I mentioned this, he laughed, and said: "Why do you think we run ten vats per cycle?"

"You get the same?" I asked, amazed.

"Yes, every tenth brew is a little off. Two are superb, most are fine, one is a bit dodgy."

"What do you do with the dodgy one?" I asked.

"We just mix it all together and no matter the variance between each vat, in the end the beer comes out tasting exactly the same. It is remarkable, but while every brew has different results when viewed per vat, when we mix the ten vats the averaging of all of them gives us the same overall taste and texture as every other brew."

I hesitate to use a pun, but does this prove Beer is Multi-Cultural? A Yeast Culture is used to brew beer, and while every batch is different, with some good some bad, they still all work together in harmony.

Editors Note: Geoff redacted all the pages he did on how to set up Commercial Brewing. I see the chapter headings, but he has removed the sheets. He was not 100% satisfied with how it worked, so set them aside. I did SEE the setup, which was fairly extensive and took up all of a garden shed. However, regrettably, the information is not available as to the process he followed.

In the last few months of his life, his tastebuds could only taste sweet. Salt and savory senses were simply gone, and with this went his desire to drink beer. He looked at me over his glasses, and said: "Imagine how much time and money I would have saved myself if I had lost my taste for booze fifty years ago!"

He is sorely missed.

BREWING TERMS

BLACK MALT or BLACK PATENT GRAIN: Malted barley grain, roasted until it is black. An essential ingredient in STOUT that gives it the characteristic color and flavor. Mat be used in small quantities to color beer

BLEACH: Household bleach. A solution of Sodium Hypochlorite that produce chlorine. A superb sterilizing agent and bottle cleaner. Must be well rinsed.

BODY: A term used to describe the solidity of texture or the consistency of beer. A beer with 'body' is the opposite of 'thin' beer.

BOTTOM FERMENTING YEAST: A Yeast which operates on the bottom of the WORT in a Closed Fermentation System. It works in the absence of oxygen. Saccharomyces Carlsbergensis is the most 'famous' of these, discovered by Carlsberg Beer. Works well in low temperatures, especially with Lager Brewing. (See Brewers Yeast" and "Wild Yeast")

BREWERS YEAST :   A single-celled fungus which ferments the sugars in the brewing process and converts them into alcohol and carbon dioxide gas. Can come in Top Fermenting or Bottom Fermenting varieties. In home brewing the yeast acts on the priming sugar used at bottling to create the 'bubbles' in the beer. A superb source of B Complex vitamins and other minerals. (Note: Yeast sold in Health Food stores as a supplement is not suitable for brewing)

BREWING: The practice of taking a mixture of malted grains (usually barley) converted to sugars, and boiling with hops and water to create a WORT. Adding yeast starts the fermentation process.

BREWING SYRUP: See "Glucose"

BOTTLE CAPPER: A device used to cap bottles. Can be as simple as the tap down variety that uses a hammer, or a more sophisticated lever action device.

BROWN ALE: An English dark ale.

BURTON-ON-TRENT: A place in England famous in brewing for its water, which contains considerable Calcium Sulphate.

CALCIUM CARBONATE: Chalk. Used to make water more alkaline. (Use one teaspoon per 5.9 US gallons)

CALCIUM SULHATE: Common Gypsum or plaster of paris. Used to harden and lower the pH of water. When mashing, it reacts with the malt and forms Phosphoric Acid. Used as three level teaspoons per 5.9 US Gallons of water.

CARAMEL: Parisian Essence, sometimes called 'burnt sugar'. Concentrated brown coloring used most often to color light malt beers.

CARAMEL MALT: See 'Crystal Malt'.

CARBONATON: The process of putting carbon dioxide gas into beer to make it sparkle. Achieved in home brewing by adding a teaspoon of sugar to the bottle when bottling. The sugar reacts with the yeast in the bottle, and forms carbon dioxide. Commercial brewers and 'pop' drink manufacturers largely add CO2 directly and do not use a fermentation process to create it.

CARBOY: A fermenting vessel, normally glass. A general name for a fermenting vessel.

CARRAGHEEN MOSS: Irish Moss. A seaweed which is boiled in the wort as a fining or clarifying agent. Used for settling protein by-products.

CEREALS: Grains. Usually barley, wheat, oats, corn, rye and rice. There is also Triticale, a hybrid of wheat and rice.

CHLORINE: A chemical used for purifying town water. This is the gas you smell when using bleach. Can be removed from water by boiling or adding a little Sodium Metabisulphite.

CITRIC ACID: An acid used to acidify water. Can be boiled with white sugar and water to make invert sugar.

CLEARING: Bottled home brew beer clears gradually after bottling as all particles in suspension settle on the bottom.

CONDITION: The bubbles of carbon dioxide gas and the 'head' of the finished beer.

BEER PACKS: The canned, pre-packaged mixture of malt and hops available in stores. The home brewer only has to add water and yeast to create a WORT, and thus beer. (see WORT)

CROWN SEAL: The cap used to seal the bottle. Usually comes with a cork or plastic liner and it forms and air-tight seal once tapped into place on the bottle.

COPPER: One of the best metals to boil up a WORT in. Old fashioned 'coppers' make excellent boilers.

CRYSTAL MALT: Malted barley grain which has been soaked in water and roasted. Does not need to be mashed to release sugars. Used for flavor and color. Sometimes called Caramel Malt.

GLUCOSE: An incomplete hydrolysis of starch. A clear syrup made from pure wheat or corn starch. A typical sample may contain 19% dextrose, 17% maltose and 64% tri, tetra and higher sugars. Adds body and contributes to the head of the beer. It is acidic with a pH of 5.2 and too much will impart a rather acid taste.

GOODS: Spent grains after mashing and sparging. Used for stock feed.

GREEN MALT: Sprouted barley grain which has not been kilned.

GROMMET: A small rubber ring used for making an airtight seal on a Fermenter.

GRAIN BREWING: Using mashed grains for brewing instead of, or together with, malt extracts.

GRAVITY: In brewing terms, the density of the WORT is measured by SPECIFIC GRAVITY. First measure if OG (original Gravity) Last measure is FG (Final Gravity).

GRIST: The mixture of malted barley and other malted r unmalted grains prior to mashing.

GUINNESS: A famous brand of stout originated by Arthur Guinness of Dublin around 1775.

GYPSUM: (see Calcium Carbonate)

HARD WATER: Water with a high mineral content.

HEAD: The frothy foam produced when beer or stout is poured.

HEADING FLUID: An additive to beer of stout to produce an artificial 'head'.

HEAD RETENTION: The ability of a beer to retain its head.

HIGHER SUGARS: The part of some sugars that cannot be fermented by yeast.

HOP EXTRACT: A highly concentrated extract from hops. 5ml or less is sufficient for 5.9 US Gallons of beer.

HOP PELLETS: Mechanically concentrated hop flowers, compressed into small pellets. 50mg is sufficient to produce 5.9 US gallons of beer.

HOPS: Flowers of the hop plant which give beer its bitterness and aroma. The main variety used in home brewing is grown in Australia under the name 'Pride of Ringwood'. There are many varieties world-wide.

HOT BREAK: The coagulation or clumping of protein matter which occurs when mashed liquor is boiled. This ultimately settles and removes some tannins.

HYRDOLYSIS: Breaking down with water.

HYDROMETER: An instrument made of hollow glass of plastic, weighted at the bottom, which when floated in a liquid gives you its Specific Gravity, or density.

HYDROMETER JAR: A tall straight-sided glass or plastic container used to hold liquid for Hydrometer readings

INFECTION: A term used for any contamination which adversely affects or ruins beer. Results from exposure of the WORT to air, or from poor sterilization.

INVERTASE: An enzyme in yeast which changes sucrose into its component simple sugars, dextrose and fructose, so tat the sucrose can be fermented. (see Invert Sugar)

INVERT SUGAR: Cane of beet sugar (Sucrose) that is boiled in water with acid to cause a separation into dextrose and fructose. This occurs naturally in fermentation by the action of the enzyme Invertase.

IRISH MOSS: (see CARRAGHEEN MOSS)

ISINGLASS: A fining agent made from fish.

KILNING: Drying and heating sprouted barley grains for malting.

KRAUSEN: The temporary rising of yeast to the top of the WORT. Krausening is more usually used as a term for a method of priming beer by adding some actively fermenting WORT to one which has finished fermenting.

LACTOSE: The sugar found in milk. It cannot be fermented by beer yeast. Sometimes used to sweeten stout, thus the old term, 'milk stout'.

LAGER: A beer fermented slowly at a low temperature using bottom fermenting yeast in a closed Fermenter, then stored for a long time at very low temperature. Derives from the German "lagern', to store.

LEES: The mixture of yeast and other fermentation by-products which settle on the bottom of the Fermenter.

LENGTH: Old English term which means total volume of the WORT prior to fermentation. A 22.5 Liter brew has a 'length' of 22.5 liters

LIQUOR: Old English term for water. Came to mean distilled alcohol at a much later date.

LITMUS PAPER: Tool used to measure pH value

LAUTER TUN: A brewery vessel for sparging mashed grains

LUPILIN: The part of the Hop Flower containing the resins and oils.

MALT: The sugars extracted from barley grain, or similar, by the conversion of its starch content. Barley is wetted and sprouted and then heated to halt the growing cycle, then dried. This changes the starch be enzyme action into a form in which it can easily be converted by further enzyme action to maltose. (during the mashing process) Barley malt is the basis for most beer, but wheat can also be used.

MALT EXTRACT: The liquor extracted from mashed barley grain, condensed into a syrup. (liquid malt extract) It is made in light and dark varieties and can also be dried into a granular powder. Diastatic liquid malt extract is condensed under vacuum to that the enzyme Diastase is not destroyed. Diastatic malt extract can be used to mash 1/5th of is weight of unmalted grains by virtue of this enzyme content. Malt extract is the basis of most commercial beer packs.

MALTOSE: The sugar in malt. $C_{12}H_{22}O_{11}$

MASHING: Heating cracked malted grain (usually barley) with water to certain temperatures for specific time periods. This converts, by the action of enzymes, the starch in the malted grain into maltose and dextrin. The general methods for mashing are known as infusion and decoction. Infusion mashing uses only one vessel for mashing. Decoction mashing means you take some of the malt, boil it in another vessel, and return it to the original mash to raise its temperature. This may be repeated once or twice.

The word decoction means to extract by boiling. Infusion means to pour a liquid in, into, or upon, also to steep r soak. With one type of infusion mashing the mash is maintained around 63 to 68C (145 to 155F) for about two hours.

MASH TUN: The vessel used for mashing

MATURATION: The period allowed after bottling or cask outing for beer to mature and improve.

MILK STOUT: Stout to which some sugar of milk (lactose) has been added to the **WORT:** It is used as a sweetener and does not ferment.

ORIGINAL GRAVITY: The Specific Gravity of a WORT before fermentation has commenced.

PITCH/PITCHING: Adding Yeast to the WORT

pH Scale: The scale from 1 to 14 that measures the acidity or alkalinity. 7 is neutral, 1 highly acid, 14 highly alkaline.

PORTER: A dark ale made with some roasted malt, similar to Stout. Reputedly made especially for London Porters in the 18th Century.

POTASSIUM METABISULPHITE: A power for sterilizing. Not generally available.

POTENTIAL ALCOHOL: The estimated alcohol by volume content the beer will have after fermentation. It is calculated from the total sugars in the WORT.

PRIMING: Putting a small quantity of sugar in a bottle at bottling time. This sugar is consumed by yeast in the bottle and creates the gassy bubbles in home brew beer. Without priming a beer will be flat and without head when poured.

PRIDE OF RINGWOOD: A variety of hop, and the principle variety grown in Australia.

PROOF SPIRIT: A method of stating the alcohol content of any beverage. This varies from country to country.

RACK: To transfer beer from one container to another, either before or after fermentation is finished. The usual purpose is to remove the beer from the by-products at the bottom of the Fermenter. Some 'rack' into glass containers so they can see the beer clearing prior to bottling. Racking is best done before fermentation is finished, so that a covering of CO_2 gas will protect the brew from infection. Only used successfully by the most experienced brewers.

ROASTED BARLEY: Malted barley grain roasted until it is black. Used to color and flavor Stout, and in small quantities in some beers. It should be cracked and boiled in the WORT.

ROLL/ROLLING: A vigorous boil.

ROUSING: Stirring the partially fermented WORT if fermentation has stopped prematurely. Not common, unless too much sugar has been used in proportion to the malt. Pumping CO_2 or Oxygen into the yeast bed to restart fermentation.

SACCHAROMYCEC: A genus of yeast which includes brewers yeast. 'Saccharo' means sugar, 'Myco' means fungus. There are top fermenting and bottom fermenting varieties.

SOFT WATER: Water without a lot of mineral content. Rain water is 'soft' and said to be preferable for making Stout and real lager.

SPARGING: Rinsing spent grains with hot water after mashing to extract all the maltose.

STARCH: The carbohydrate content of grains.

STARCH END POINT: The point at which all starch in grain being mashed as been converted to sugars. (Test by placing 1 teaspoon of liquor on a white surface and add 1 drop of tincture of iodine. If it changes to blue, starch is still present in the mash)

STARCH HAZE: A cloudiness consisting of nitrogenous compounds from proteins in the grain which can appear in beer. Occurs most often after mashing grains and bottling without the use of finings.

STERILIZING: Treating equipment with chemical, bleaches or boiling to kill bacteria. The most essential process is all forms of brewing and the most important word in this book.

STOUT: Beer to which black patent grain or roasted barley has been added when boiling the WORT. This gives Stout its characteristic look and flavor. It has no nutritional value over ordinary beer.

STUCK FERMENT: (see Rousing) A ferment that has prematurely ceased.

STRIKING TEMPERATURE: In mashing, the temperature of the water before the grain is added.

SWEET WORT: The liquor from mashed and sparged grain before added hops.

THERMOMETER: An instrument for measuring temperature.

THIN: Beer that lacks Body.

TORREFIED BARLEY: Barley grain roasted like popcorn. Used as a flavoring agent.

TUN: A vessel in a brewery

UNFERMENTABLE SUGAR: Any sugar which beer yeast will not ferment, such as higher sugars and lactose.

VORLAUFING: German for "Recirculation". When grain mash is used, this is a clarifying process to strain particles from the WORT.

WHEAT MALT: Malt made from wheat, instead of barley.

WILD YEASTS: Unclassified yeast organisms that are always present in the air. They must be kept out of the WORT.

WORT: The full volume of liquid to be fermented when all water has been added. The mixture before yeast is added.

YEAST: See Brewers Yeast

ZYMASE: The group of enzymes in yeast which causes fermentation and breaks down sugars into alcohol and carbon dioxide gas. Without this enzyme there would be no such thing as beer.

Final Note: *Geoff Wallace was a perfectionist when it came to beer. It was his passion, his curiosity and something that went beyond a hobby. In his files there are letters to Universities, breweries, all kinds of places where information might be had, all asking about some specific point that might have only had a passing mention in this book.*

From his experiments and from what he gleaned from the experts, he compiled a depth of knowledge about the art of brewing that was second to none. In these pages is what you would call the "bare bones", but it is more than enough to give you a sound foundation upon which to build your brewing career.

To close, I would like to share with you one of my favorite stories about my father, the thing that led him to become extremely rich. He was back in Sydney after the World War Two, and working as a Bar Keep at the Coogee Bay Hotel. As always, it was a way of keeping body and soul connected, but then a ray of light appeared!

Into that very bar walked a bunch of his Navy friends. What else was Geoff to do but forget his duties and sit to drink with his mates. Other men may have been more concerned about what the boss might have thought, but Geoff was a free spirit who loved a beer, and loved a laugh. Plus, he was never very good with consequences.

Obviously, he was fired as soon as the boss walked in and saw what was happening, but as a result, he took up an offer of a position on the wharves in New Guinea, which in turn led to buying a plantation, which in turn led to him making a fortune. He passed away with full faculties at age ninety one.

So this is unequivocal proof that BEER, when drunk in sufficient quantities, will lead to wealth, health and happiness.

Here's looking at you, Geoff!

Want more?

The biography of Geoff Wallace is available on Amazon: goo.gl/mGSHwn

Photos of the Author from his time in the Navy, to holding the Editor on his knee, to the orange haired hipple. (It's a wig - he turned up in it one day for a joke)

See You Later Geoff

FERMENTER

HOPS and BARLEY

THOMAS
COOPERS
MALT EXTRACT
100% PREMIUM MALT
NET WEIGHT 1.5kg ℮ (3.3.lb) – 1.(1.(3))
DARK

HOP FIELD

YEAST

HOP PELLET

For over 13,000 years man has been brewing BEER now YOU can too!

BEER, all beer, not just Home Brew, is a mixture of MALT, HOPS, YEAST and WATER. Sugar is used to increase the alcohol content, and temperature is controlled to assist fermentation.

Everything you need is in this book.

PARABLES OF GEOFF

STORIES AND ANECDOTES OF
GEOFFREY JOHN WALLACE

A Biography by Michael Wallace

A BIOGRAPHY FOR THE FRIENDS AND FAMILY
OF GEOFF WALLACE, AND ANYONE ELSE WHO
IS INTERESTED IN READING HIS TALE

Want more? The biography of the writer of this
book is up on Amazon: goo.gl/mGSHwn

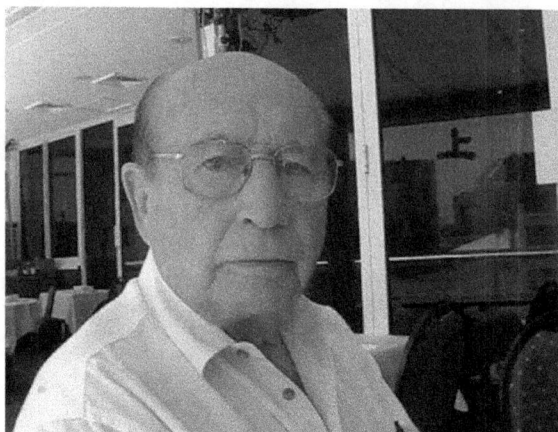

Geoff Wallace, outside his favourite Yum Cha
restaurant at Broadbeach in 2012

BEER O'CLOCK
COPYRIGHT 2018 Michael Wallace

ISBN: 978-0-6484277-0-4
Copyright 2018 Michael Wallace
Publisher: Ladder to the Moon Productions
Email: info.numberharmonics@gmail.com
Web: laddertothemoon.com.au

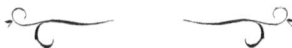

www.ingramcontent.com/pod-product-compliance
Lightning Source LLC
Chambersburg PA
CBHW071237090426
42736CB00014B/3121